Decision-Making for Biomass-Based Production Chains

The Basic Concepts and Methodologies

Şebnem Yılmaz Balaman
Department of Industrial Engineering, Dokuz Eylül University,
İzmir, Turkey

ELSEVIER

ACADEMIC PRESS
An imprint of Elsevier

Academic Press is an imprint of Elsevier
125 London Wall, London EC2Y 5AS, United Kingdom
525 B Street, Suite 1650, San Diego, CA 92101, United States
50 Hampshire Street, 5th Floor, Cambridge, MA 02139, United States
The Boulevard, Langford Lane, Kidlington, Oxford OX5 1GB, United Kingdom

British Library Cataloguing-in-Publication Data
A catalogue record for this book is available from the British Library

Library of Congress Cataloging-in-Publication Data
A catalog record for this book is available from the Library of Congress

ISBN: 978-0-12-814278-3

Publisher: Brian Romer
Acquisition Editor: Raquel Zanol
Editorial Project Manager: Naomi Robertson
Production Project Manager: Bharatwaj Varatharajan
Cover Designer: Matthew Limbert

Typeset by MPS Limited, Chennai, India

Contents

Preface ix

1. Introduction to Biomass—Resources, Production, Harvesting, Collection, and Storage 1

1.1 Biomass Concept and the Main Biomass Resources 1
1.1.1 Dedicated Energy Crops 2
1.1.2 Residues 8
1.1.3 Wastes 13
1.2 Recent Trends in Biomass Production 19
1.3 Discussion 20
References 22
Further Reading 23

2. Biomass-Based Production Systems 25

2.1 The Main Systems, Technologies, and Processes for Biomass-Based Production 25
2.1.1 Thermochemical Conversion Technologies 26
2.1.2 Biochemical Conversion Technologies 41
2.1.3 Biomass Extraction 49
2.2 Recent Trends in Biomass-Based Production Technologies 51
2.3 Discussion 53
References 54

3. Biomass-Based Supply Chains and Logistics Networks 55

3.1 General Structures and Key Components of Biomass-Based Supply Chains 55
3.1.1 Biomass Cultivation, Harvesting, Collection, and Storage 58
3.1.2 Biomass Pretreatment 59
3.1.3 Biomass Transportation 59
3.1.4 Biomass Conversion 67
3.1.5 Bio-Fuel Storage, Transportation, and Consumption 67
3.1.6 Energy Production and Transmission 72
3.2 Discussion 73
References 74

4. Sustainability Issues in Biomass-Based Production Chains 77

4.1 Sustainability of Biomass-Based Production Chains 77
4.1.1 Economic Sustainability 78
4.1.2 Environmental Sustainability 82
4.1.3 Social Sustainability 86
4.1.4 Technical Sustainability 89
4.2 Design, Planning, and Implementation of Sustainable Biomass-Based Production Chains 91
4.2.1 Sustainability Indicators, Standards, and Certification Schemes 91
4.2.2 Sustainability Assessment in Design and Management of Biomass-Based Production Chains 100
4.2.3 Life Cycle Assessment of Biomass-Based Production Systems 105
4.3 Discussion 109
References 111
Further Reading 112

5. Uncertainty Issues in Biomass-Based Production Chains 113

5.1 Uncertainty Sources in Biomass-Based Production Chains 113
5.1.1 Uncertainty in Market Demand and Prices 114
5.1.2 Uncertainty in Supply of Biomass and Other Materials 116
5.1.3 Technology and Process Uncertainties 117
5.1.4 Uncertainty in Governmental and Regulatory Policies 119
5.1.5 Uncertainty in Transportation and Logistics 119
5.1.6 Natural Uncertainties and Disruption Possibility 120
5.2 Handling Uncertainties in Biomass-Based Production Chains 121
5.2.1 Approaches for Uncertainty Modeling 122
5.2.2 Methods for Uncertainty Impact Evaluation 137
5.3 Discussion 139
References 141

6. Basics of Decision-Making in Design and Management of Biomass-Based Production Chains 143

6.1 Decision Levels in Design and Management of Biomass-Based Production Chains 143
6.1.1 Strategic Decisions 144
6.1.2 Tactical Decisions 149
6.1.3 Operational Decisions 151
6.2 Decision-Making in Design and Management of Biomass-Based Production Chains 153
6.2.1 The Main Objectives and Criteria 154

6.2.2 Multiple Attribute and Multiple Criteria Decision-Making Methods in Biomass-Based Production Chains 165
6.3 Discussion 178
References 179
Further Reading 183

7. Modeling and Optimization Approaches in Design and Management of Biomass-Based Production Chains 185

7.1 Mathematical Modeling-Based Approaches 185
7.1.1 Linear Programming 186
7.1.2 Nonlinear Programming 192
7.1.3 Methods and Tools to Solve Linear and Nonlinear Programming Problems 194
7.2 Simulation-Based Approaches 206
7.3 Artificial Intelligence-Based Approaches 213
7.3.1 Computational Search and Optimization Methods 213
7.3.2 Machine Learning 226
7.4 Geographic Information Systems 230
7.5 Game Theory 231
7.6 Discussion 234
References 235
Further Reading 236

Index 237

Preface

The ever increasing attention in many parts of the world on three significant issues has leveraged the use of biomass sources in a variety of production processes to generate final products with critical importance for human life, such as electrical and thermal energy, fuel, chemicals, as well as useful by-products (e.g., bio-char and digestate); global warming, environmental pollution, and the depletion of nonrenewable sources. In parallel with this attention, researchers and professionals from different areas (e.g., governmental units, universities, private companies) have particularly concentrated on design, planning, operation and management of biomass-based production systems and biomass-based supply chains, with a specific focus on bioenergy production, which has also been promoted by the European Union. However, as biomass-based production is an emerging field, the main challenge is the development and application of reliable and effective decision-making methodologies to handle and solve a variety of problems in biomass-based production chains with different activities and operations, (from cultivation, harvesting, transportation, storage to conversion, consumption, and by-product use) stakeholders, and stages.

How far biomass-based production can further contribute to the solution of the economic, environmental, and social problems raised by abovementioned three vital issues, strongly depends on the cost and availability of sustainable biomass sources and production technologies. Secure, long-term supply of sustainable feedstock, as well as reliable and efficient technologies and processes, are essential to the economics of biomass-based production. In addition, given various origins and quite different physical properties of biomass and a wide range of bio-products to meet the consumer requirements, many different alternatives for technologies, systems, and processes have been introduced to convert biomass to the final products. For these reasons, worldwide research and innovation support is directed at reducing the feedstock and technology costs, diversifying the feedstock alternatives, and increasing the overall efficiency of the conversion technologies.

To accelerate the innovation and competitiveness in biomass-based production systems, considerable progress has been made to ensure efficiency and productivity in biomass-based production chains besides stable regulatory schemes in the final product markets. One of the core priorities, to boost the research and innovation in biomass-based production, is to develop integrated frameworks to design resilient and responsive supply chains, assign efficient and timely operations in all stages of the chain, and ensure the continuity and sustainability of these operations. Concentrated efforts on developing new techniques and further improvement of current methodologies, through innovative integration of different approaches based on operations research and production management principles, with particular emphasis on sustainability and uncertainty issues, are essential for design, planning and optimization of biomass-based production chains. In addition, these methodologies should be reliably applied to cope with varying problems beginning from the phase of biomass procurement to the production and distribution of final products and by-products with the overall focus of the targets on cost reduction, improvement of conversion efficiency, and on meeting sustainability goals. Another challenge in biomass-based planning problems is that, due to the complex trade-offs involved in conflicting criteria and objectives, various competing design decisions that affect the system performance cannot be made independently. Besides the incorporation of inherent uncertainties, multiple (usually conflicting) objectives to reflect different aspects of sustainability in biomass-based production chain design models that emphasize economical, technical, environmental, and social issues, have to be considered in real-life problems, simultaneously. It is also importantly problematic to develop or use a decision-making procedure that recognizes the links between the strategic, tactical, and operational decision levels, as well as providing optimum decisions at all levels through appropriate qualitative and quantitative approaches, to create and maintain the competitive advantage. Understanding the main activities, systems and processes at all these decision levels and having the ability to link these activities and processes are of critical importance in enhancing building and employing appropriate analytic and quantitative decision-making approaches in design, management, and operation of biomass-based production chains. This book provides all the information required to comprehend the activities and processes at all decision levels, as well as the modeling and solution frameworks.

The objective of this book is to provide a comprehensive knowledge on a wide range of topics that a researcher or practitioner, dealing with decision-making in biomass-based production chains, will need to know to develop efficient methodologies for strategic, operational, and tactical decision-making and in applying the most appropriate technique(s) to solve variable economic, environmental, and social issues. The book mainly comprises two parts. The first part of this book (Chapters 1–5) introduces:

- biomass resources and associated harvesting, collection, storage, and distribution systems for those sources,
- technical and technological background of biomass-based production systems,

- all stages and related activities in biomass-based supply chains and logistics networks,
- and finally, the uncertainty and sustainability issues in biomass-based production chains along with the methods for incorporating these issues in problem-solving procedure.

The second part (Chapters 6 and 7) looks at:

- the fundamentals of decision-making in design, management, and operation of biomass-based production chains describing strategic, tactical, and operational decision levels and related decision problems at all levels,
- the main objectives and criteria used in modeling and solving decision problems in biomass-based production chains,
- the multiple attribute and multiple objective decision-making methodologies,
- a wide range of modeling and optimization methods and varying techniques to find the optimum solution or to approximate the optimum solution of the decision problems in biomass-based production chains.

This book allows readers to clearly identify the decision problems in biomass-based production chains, and find the most appropriate modeling and solution methodologies, to meet the problem-specific requirements effectively. This book would make an excellent textbook for coursework or self-study on this living and rapidly changing topic. Students, researchers, and working and practicing professionals in the fields of biomass, bio-energy, renewable sources, renewable energy, sustainable supply chains, environmental management, and sustainability, can utilize this book. It also serves as a guide to the nonexpert who desires to have an overview of this industry, which has profound effects on agricultural and energy markets around the world. Providing a detailed perspective on the biomass sources, technologies, supply chain stages, main problems, decision levels, and main criteria/objectives, the book supports the researchers and decision-makers in developing methodologies for emerging problems, considering problem-specific requirements, by linking the strategic, tactical, and operational decision-making levels, and taking into account the most important criteria/objectives. The book also introduces the modeling and solution frameworks, giving comprehensive information on these frameworks to provide a detailed perspective, which methodologies the readers can utilize in developing new approaches and solving the problems in design, management, and operation of biomass-based production chains.

I remember how much I needed a comprehensive textbook comprising all the basic concepts about biomass-based production systems and supply chains during my PhD studies. This was my first and strongest motivation to write this book. I realize now how much patience was needed from my friends and colleagues, who encouraged me to write this book, which took me much longer than I anticipated. I am grateful for their support. I am also deeply grateful to my mother Mürüvet Yılmaz and my family for their endless support and belief in me. I appreciate the support and cooperation I have received from to the editorial staff of Elsevier, Raquel Zanol, Sabrina Webber and Naomi Robertson. And, my special thanks go to my husband, Olcay Balaman for his unwavering support and understanding during my studies, and to my beloved son Mert, to whom I dedicate this book.

Şebnem Yılmaz Balaman

Chapter 1

Introduction to Biomass—Resources, Production, Harvesting, Collection, and Storage

Chapter Outline

1.1 Biomass Concept and the Main Biomass Resources 1
1.1.1 Dedicated Energy Crops 2
1.1.2 Residues 8
1.1.3 Wastes 13
1.2 Recent Trends in Biomass Production 19
1.3 Discussion 20
References 22
Further Reading 23

1.1 BIOMASS CONCEPT AND THE MAIN BIOMASS RESOURCES

Biomass is a biological material derived from living, or recently living organisms available on a renewable basis, which contains complex mix of carbon, nitrogen, hydrogen, and oxygen. Biomass resources, sometimes referred to as bio-renewable resources, are all forms of organic materials, including plant matter both living and in waste form, as well as animal matter and their waste products. Various biomass resources are converted into bio-products by different conversion processes in biomass-based production chains throughout the world. But the key to utilizing biomass resources in a beneficial way is to focus on the most appropriate resources, considering the characteristics of the conversion and handling systems, and use them at an appropriate scale. A good understanding on biomass resources as well as their production, collection, handling, pretreatmenttreatment, and storage is critical for efficient design and management of biomass-based production chains. This chapter presents a comprehensive overview on biomass resources, as well as production, harvesting, collection, handling, pretreatmenttreatment, and storage processes, for all overviewed biomass resources under three categories of biomass, energy crops, residues, and wastes. Fig. 1.1 depicts the biomass resources focused in this chapter.

Decision-Making for Biomass-Based Production Chains. DOI: https://doi.org/10.1016/B978-0-12-814278-3.00001-7

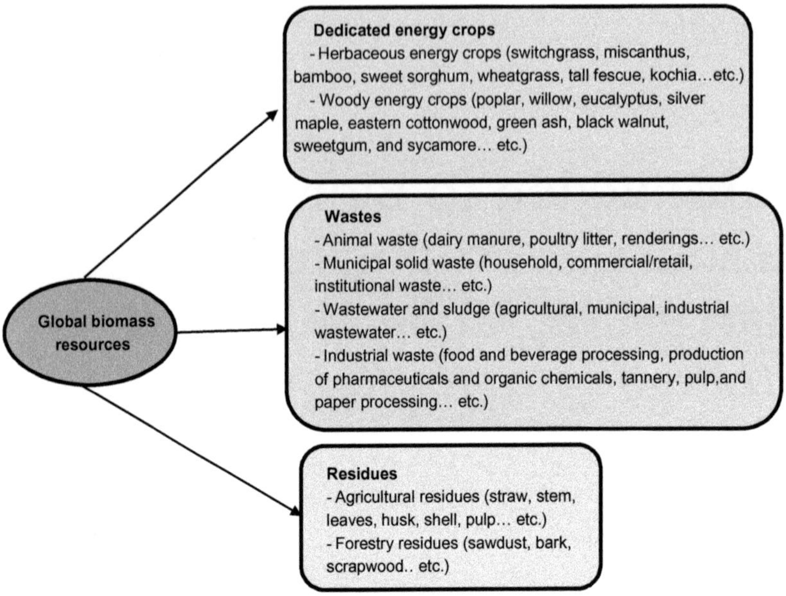

FIGURE 1.1 The classification of biomass resources covered in this chapter.

1.1.1 Dedicated Energy Crops

Until the first oil crisis of the 1970s, when a sudden rise in the price of oil led to the first push for the development of renewable energy production, energy crops have been largely ignored in favor of growing food crops to feed people and animals. After this period, many governments supported the development of novel nonfood crops for energy production in addition to food production. Interest in crop production for energy purposes is now increasing as the economics of extracting fossil fuel-based energy becomes more expensive and there is an increasing concern over energy security. In addition to volatile and increasing oil prices, numerous reports draw attention to the substantial cost to humankind of not acting to reduce the current rate of increase in greenhouse gas emissions. As the use of fossil fuels significantly contributes to climate change, the need for alternative energy sources to save carbon is placed at the top of the global agenda. However, the global population continues to grow at an alarming rate, and people both need to be fed and are consuming more energy. This raises trade-offs and questions of "Food versus Fuel," how much land and other resources are available, what are the priorities, and how should they be divided between food and fuel production? Over the years, the large-scale production of energy crops has become a highly controversial topic, in that each crop must compete for arable farmland for food purposes, and as to whether or not it is sustainable or viable to use such large areas of farmland and forests, to produce dedicated energy crops, has been the subject of much debate. These questions should be answered and trade-offs should be considered while designing and managing biomass-based production systems and supply chains, especially in planning the land-use for energy crop production.

Dedicated energy crops are defined as nonfood energy crops that are unsuitable for human or animal consumption, and are grown primarily for the purpose of producing biomass, to be converted into energy and/or fuel on marginal land dedicated to provide biomass. Dedicated energy crops can be investigated in two general categories; herbaceous energy crops and woody energy crops.

1.1.1.1 Herbaceous Energy Crops

Herbaceous energy crops, also referred to as grassy or forage energy crops, are high-yielding lignocellulosic crops that remain in cultivation for several seasons and are harvested, on average once a year, after taking two to three years to reach full productivity. Energy crops within this definition are perennial in nature, so that they can be cut and harvested for biomass over successive years without recultivation or sowing, and the whole crop can be harvested and used for energy production.

They have several advantages over other types of energy crops, such as, they have efficient solar energy conversion resulting in high yields, need low agrochemical inputs, have a low nutrient and water requirement due to their extensive rooting system, which holds onto fertilizers and water, have low moisture levels at harvest, and plants with perennial growth habits also have the benefits of low establishment costs and fewer annual operations (Fernando et al., 2017; Adams and Lindegaard, 2016). These include grasses such as switchgrass, miscanthus (also known as elephant grass or e-grass), bamboo, sweet sorghum, wheatgrass, tall fescue, kochia, etc. Among them, switchgrass and miscanthus are the most commonly used biomass resources because they are adaptable to many climatic zones, require relatively low water and nutrients, are adaptable to low-quality land, and have a minimum of 10 years productive life (10 years for mischantus, 10–20 years for switchgrass) (Allen et al., 2014). Also, they have positive environmental impacts because of their deep roots and ground cover, so that the cultivation of them in marginal soils has the potential to restore soil properties (fertility, structure, organic matter) and reduce erosion (Stewart et al., 2015).

1.1.1.2 Woody Energy Crops

Woody energy crops (also referred as short-rotation woody crops), mainly consist of fast-growing hardwood tree species that are harvested within five to eight years of planting. These include poplar, willow, eucalyptus, silver maple, eastern cottonwood, green ash, black walnut, sweetgum, and sycamore. Among them, poplar and willow are also referred as Short-Rotation Coppice.

Woody energy crops feature high biomass productivity, that is, higher than long-rotation forest systems grown on forest land (Hauk et al., 2017). They also require a low level of input when compared with annual crops. In addition, the production of woody energy crops requires few working steps and little to no pesticide and fertilizer; hence the production of biomass is energy efficient (Don et al., 2012). They also have positive external effects, such as reduced soil erosion and increased soil quality. Despite these benefits, they are not widespread in North America and many European states, such as in southern Germany, due to long investment periods and perceived economic disadvantages (low profitability and high risk) (Don et al., 2012).

Short-rotation woody crops can be specified as purpose-grown trees that are planted as purpose-grown wood on sites that enable high productivity and proximity to the processing plant. These purpose-grown trees offer a variety of inherent logistical benefits and economic advantages, compared with other lignocellulosic energy crops, because of their ability to provide a stable supply of feedstock and relatively lower collection and handling costs (Hinchee et al., 2009). The fact that trees provide a renewable inventory of available biomass through year-round harvest and a continuity of growing, year by year, results in a flexibility in harvest time that leads to advantages, such as minimized inventory holding costs, reduced degradation losses associated with storage of annually harvested biomass, as well as production, distribution, and storage throughout the year ensuring full capacity utilization at a processing plant. In addition, the flexibility that trees can be harvested at different times aids in accommodating annual fluctuations in biomass supply due to natural reasons (e.g., disease, pest, or drought).

1.1.1.3 Plantation and Harvesting of Dedicated Energy Crops

1.1.1.3.1 Site Selection and Preplanting Site Preparation

Since herbaceous energy crops are perennial, and that they will exist on the dedicated plantation site for at least 15–20 years, site selection is an important phase of designing supply chains for the production of herbaceous energy crops. These crops can reach up to 3.5 m in height, hence their impact on the local landscape, particularly if the site is close to a footpath or an adjacent landowner, needs to be considered when selecting the plantation site.

Currently, the leading countries in areas planted for energy generation from Short Rotation Coppice (SRC) are Sweden and the United Kingdom (Mola-Yudego and González-Olabarria, 2010). SRC can be planted on a wide range of soil types from heavy clay to sand, including land reclaimed from gravel extraction and colliery spoil (DEFRA, 2007). As SRC requires more water for its growth in comparison with any other conventional agricultural crop, the plantation site should have a good moisture-retentive soil. The area of the site should be adequate (large and regular shaped) for large harvesting machinery is involved. At the end of a three-year growing period, the SRC will be up to 8 m tall, hence creating a three-dimensional mass in the landscape. For this reason, it is important to design and construct SRC plantations to prevent adverse views on the landscape.

It is essential to establish the plantation site correctly and effectively, to avoid possible future problems, since the crop has the potential to be in the ground for at least 15 years. For effective establishment and crop management to guarantee high yields, thorough site preparation is essential.

As the first step in the site preparation for herbaceous energy crop plantation, the site is sprayed with a suitable herbicide, to control perennial weeds in the autumn before planting and the site should be cultivated immediately before planting in the following spring. Therefore, soil aeration will be improved and root development will be supported to enhance the establishment (Planting and Growing Miscanthus, Best Practice Guidelines, 2007). For the case of SRC plantation site preparation, at least 10 days before plowing the SRC plantation site, herbicide application is required.

1.1.1.3.2 Planting and Harvesting

Herbaceous energy crops are established from seed, which is generally available from commercial suppliers. First-year growth is insufficient to be economically worth harvesting, hence, after planting, a minimum of two to three years is required to achieve maximum annual yields. Weed control is important during the first year (Wolf and Fiske, 1995). Although irrigation is necessary to grow switchgrass and other warm-season plants, permanent irrigation may not be cost-effective for herbaceous energy crop systems. The deep and extensive root system of herbaceous energy crops enables them to sustain good growth even in periods of infrequent rainfall. Despite this fact, Ocumpaugh et al. (2003) stated that a watering interval of seven days or less was critical to obtaining seedling survival of 90% or more in all soils.

Underground storage organs (also known as rhizomes) of Miscanthus, of which pieces can be replanted to produce new plants, naturally provide slow spreading. Miscanthus is quite an adaptable energy crop by means of its roots that can extract water to a depth of around 2 m, and it grows on a range of soils, from sand to high organic matter soil. However, the yields are strongly influenced by annual rainfall frequency and soil water retention at any site. To obtain a high yield of herbaceous energy crops, key determinants are sunshine, temperature, and water availability, of which annual variability results in annual yield variations. Also, any site-specific microclimatic conditions will affect the annual yield.

The key determinants that influence SRC yield are water availability, weed control, light, and temperature. The root system and stumps of SRC are well-established where the nutrients are able to be stored to guarantee vigorous growth for the shoots. The high moisture content of the soil in the spring and the adequate amount of sunshine in the summer make March an ideal planting time for SRC. The large initial investment cost of plantations may be intimidating since there is no financial payback for four years from the initial establishment. However, in the United Kingdom grants are available to support establishment and in Sweden an extensive scheme of subsidies was developed between 1991 and 1996, being reduced after that time (Mola-Yudego and González-Olabarria, 2010). Although SRC is a relatively low-return crop, it requires low-maintenance, and planting SRC is a way of utilizing difficult and idle fields with little need for pesticides or treatments.

Harvesting of herbaceous energy crops has usually been accomplished using conventional mowing, raking, and baling equipment on slopes of less than 35 degrees (Wright et al., 2011). The timing of the harvest depends on the primary aim of crop-use and may occur once or twice during the growing season, that is, twice if harvested for forage, once if harvested for bio-energy production. Harvesting can take place during the winter or from late February to early May, as long as weather conditions permit. However, delayed harvests may cause some loss of biomass. To obtain a dry and baled product, which is desirable for energy generation purposes, the crop is cut with a forage harvester, then baled with a number of different types of baler, each suitable for different scales of energy production facilities and producing bales in different shapes (e.g., rectangular, round, and compact rolls).

Harvesting of SRC takes place on a two- to five-year cycle, and are carried out in winter when the soil is frozen (usually from December to mid-March). There are four different types of harvesting SRC; direct-chip harvesting, harvesting as entire rods, harvesting into round bales, and harvesting into billets (Caslin et al., 2010). To store the SRC more easily, shoots are harvested as whole stems, which then can be dried until the next autumn when the moisture content of the wood will decrease to about 30% on average. Then, the dried stems can be either converted into wood chip or cut further into billets depending on use. In the case of wood chip being produced, the most efficient way of harvesting is to use direct-chip harvesters that cut and chip the shoots on a loading platform. Direct chipping in the field is less costly than separate chipping in the store, however, after direct chipping, the wood chip should be well stored to prevent degradation and composting.

Single-pass cutting, bundling, cutting–baling, or cutting–chipping equipment can be used for Short Rotation Willow (SRW) harvest-to-delivery logistics as well as two-pass cutting–baling or cutting–bundling systems. Cutting–baling or cutting–chipping systems are more commonly used in North America, while cutting–bundling equipment is widely used in Europe. SRW harvest-to-delivery can use the same equipment as that used for harvesting and transporting understory forest biomass feedstock. Cutting, baling, and handling systems for SRW coppice usually consume more energy than those for the energy grasses (Miao et al., 2012).

1.1.1.4 Treatment and Storage of Dedicated Energy Crops

Herbaceous energy crops are stored as bales and the storage procedure generally follows the same rules as the handling and stacking of any bales produced in agriculture, which are specified by the governmental units in each country. For example, in the United Kingdom, these rules are explained in "Handling and Stacking Bales in Agriculture" by the Health and Safety Executive. Health and safety risks can be reduced and effective supply logistics can be ensured by correctly siting the bale stacks in the storage area. As stacks are combustible, they should be located away from public roads, footpaths, to reduce the risk of fire from discarded cigarette ends and sparks, and from overhead power lines.

The grass, which is usually collected in large round or rectangular bales, is moved to the field edge for storage. High moisture content makes nutrients in the grass more accessible to fungi and bacteria, which cause material degradation and loss; hence moisture reduction in the harvested grass is necessary depending on the storage time required. Long-term storage needs moisture levels below 20% for storage as bales and this can be attained with field drying as the harvested crop has higher moisture content. Covering the bales is also important to limit degradation and biomass losses, and to keep them dry. Plastic sheeting or bale stack sheets can be used for covering, which are available from agricultural suppliers. Bale stack sheets are stronger and easy to secure using the attached guy ropes (Planting and Growing Miscanthus, Best Practice Guidelines, 2007).

Storage for short periods in the year adds no additional cost except for handling. However, in some cases long-term storage may be required, which options for bales include (1) storage outside with no protection, (2) storage outside on a gravel pad with a tarp cover, and (3) storage inside a building (Turhollow et al., 2009). Herbaceous energy crops can also be stored as silage or as wet piles. In these cases, the mown grass is harvested with a silage chopper, rather than a baler. The best drying and storage regime is dependent on the climate patterns at the crop production site and the intended end use (Wright et al., 2011).

Freshly harvested woody energy crops have a minimum moisture content of 50% and they need to be dried before storage to prevent decomposition and material losses. Besides the needed storage time, the need and dosage of drying will be specified considering the harvesting system used, that is, direct harvested chip requires immediate drying and dedicated drying facilities, whereas whole rods and billets will dry naturally. Drying in dedicated drying facilities is a costly operation because of the excess heat that is required. However, when natural drying is applied to the whole rods and billets, moisture levels of 30%–25% can be attained, which is adequate for only very short-term storage when the rods/billets are chipped on demand.

1.1.2 Residues

1.1.2.1 Agricultural Residues

Agricultural residues are the nonedible portion of plants that remain on fields following harvesting, and which are differentiated from other biomass resources (e.g., energy crops), as they are in most cases not specifically produced for energy purposes (Lal, 2004). The European Committee describes agricultural residues as "biomass residues originating from production, harvesting, and processing in farm areas." Leftovers that are generated through harvesting of crops or that are discarded during crop processing into the generic category of crop residue can also be included into this category (Lal, 2004).

Agricultural (or crop) residues comprise all residues from harvesting activities, such as straw, stem, stalk, leaves, husk, shell, peel, pulp, stubble, etc. These residues can be obtained from cereals (rice, wheat, maize or corn, sorghum, barley, millet), cotton, groundnut, jute, legumes (tomato, bean, soy), coffee, cacao, tea, fruits (banana, mango, coco, cashew), and palm oil. From rice, both straw and rice husks are produced, which are efficient feedstocks for the energy conversion process. Also, significant quantities of biomass remain in the fields in the form of cob when maize is harvested, which can be converted into energy. Sugar cane harvesting leads to harvest residues in the fields, while processing produces fibrous bagasse, both of which are good sources of energy as well. All these lignocellulosic materials can be converted into bio-products (bio-energy, bio-fuel, bio-fertilizer, etc.) by a wide range of conversion technologies. Since agricultural residues are by-products of food production systems, the biggest advantage of utilizing agricultural residues in biomass-based production is that they do not compete with food production and they are perceived to be sustainable sources of biomass for energy that do not threaten food security.

Due to the large annual variation in agricultural crop production yields, the amount of agricultural crop residue varies significantly. Crop production yields are influenced by the cultivated area, types of crops grown, specific climate and soil conditions, water availability, sowing rate and sowing date, and management practices (tillage, density of planting, fungicide treatment, harvesting and collection method, fertilization, etc.). The yield of residue obtained from crops is highly dependent on the residue to crop ratio, which describes the relationship between the biomass that grows as the main product and the residue from it. This ratio is directly relevant to the portion of biomass technically harvestable and the cutting height. Estimating this ratio accurately is usually difficult as it is significantly impacted by uncertainty in climate and soil conditions as well as agricultural practices. The quality of residue in terms of energy production potentials or with regard to storage and transportation demands varies between crops. The other factors that impact the yield of residue are the cultivated area, crop yield, type of crop,

crop rotation, plant variety, harvesting techniques, and management practices including harvesting and logistics (Monforti et al., 2013). Therefore, the factors affecting residue yield can be grouped into those that are noncontrollable, that is, site conditions and climate, and two main controllable categories, (1) crop breeding and (2) management options including harvesting and logistics (Iqbal et al., 2016). Breeding has direct effects both on crop yield and on residue to crop ratio. Residue to crop ratio is affected by breeding programs because the primary aim of these programs is to maximize the yield of main food/feed grain by increasing plant allocation to the product for a high yield. This option also requires avoiding bending of stalks or whole plants, which leads to a decrease in straw length, and even among cultivars with a high grain yield, significant differences in straw to grain ratio can be observed. Briefly, there are two ways to increase residue yield, (1) increasing actual crop yield and (2) increasing the residue to crop ratio by improving the residue recovery and collection rate through use of appropriate harvesting techniques. The focus of the first part is mainly on the choice of variety type, crop management measures and specific adaptations, and different ecological and climatic conditions, whereas the second part mainly deals with using the best harvesting procedures and technologies, optimizing the rate of residue recovery considering crop and region specifications, and finding the best ways of transport, storage, and handling of residues.

There are other competitive uses of straw such as, incorporating crop residues in the soil as a precaution against soil erosion, as a fertilizer, and soil structure improver, using the straw for crop protection in cold climates when crops are left in the ground during winter, utilizing straw to produce pulp and paper or as insulating materials for buildings, and the use of straw for livestock feeding and bedding, which are the most important competitive uses of straw. Kretschmer et al. (2013) stated that the competing existing uses of straw are key barriers that affect the current functioning of the straw supply chains. In addition, all residues cannot be available due to residue retention and loss during transportation and storage, and subsequent processing. Most studies assume that about 25% of the total available agricultural residues can be recovered (Hoogwijk et al., 2003). According to Panoutsou et al. (2009), from the total agricultural residues produced in Western European Countries, 48% are being exploited in non-energy (e.g., animal feeding) or traditional energy applications and a further 40%–45% of the unexploited quantity cannot be exploited for various technical and/or economic reasons (Panoutsou et al., 2009). These factors should be taken into consideration when estimating the maximum available supply of residue biomass.

1.1.2.2 Forestry Residues

Forestry residue is one of the major sources of biomass-based production chains, which include biomass material not harvested or removed from logging sites in commercial hardwood and softwood stands for manufacturing purposes, such as extracting stem-wood for timber or pulp production, as well as material resulting from forest management operations, such as cleaning and thinning young stands, cutting and felling older stands, logging for clearing roads, and natural attrition (being damaged by insects, disease, or fire, etc.). Excess small pole trees, rough or rotten dead wood, and other woody material, which can be collected after a harvest, can be used as biomass resources from forestry operations. There are two broad categories of forestry residues – primary forestry residues produced when harvesting timber, and secondary forestry residues produced during the processing of biomass-based materials, such as sawdust, bark, and scrap wood. The second category is also referred to as biomass processing residue. For example, the processing of wood for products or pulp produces unused sawdust, bark, branches, and leaves/needles. All of these residues have alternative uses, that is, they can then be combusted for heat and energy production, or converted into additional bio-products, that is, wood pellets can be generated from diverting forestry residues, to be used for the energy or paper industry, which will result in additional biomass extraction. It should be noted that, encouraging use of unused primary forestry residues instead of leaving them on the land may lead to loss of organic matter, soil carbon, and habitat for biodiversity.

In typical practices, only timber of a certain quality can be used in lumber mills and other processing facilities. Also, some parts of the forest may not be available for felling due to protection or preservation policies, and geographical challenges may prevent easy access to much of the forest. As a result of these factors, only a certain proportion of any forest will be available for harvesting for the purpose of supplying feedstock for the timber, pulp, and paper industries. According to the Food and Agriculture Organization of the United Nations, less than two-thirds of a typical harvested tree is taken from the forest for further processing to produce wood products, that is, harvesting operations for timber production usually remove only 25%–50% of the total volume and the remaining part can be used for the biomass-based production industry, if they are not left in the forest, disposed of through open burning via forest management programs, or collected for local fuel wood. The whole residue part of the harvested tree cannot be recovered due to various losses. The remaining unused parts that can technically be recovered comprise branches and tops, above ground stumps, sawdust, fines, and bark. Twigs and foliage are generally deducted from the top and branches of a tree since they can cause problems further down the supply chain for wood chip sizing and pellet manufacturing.

The residue generation processes can be divided into three broad sections (Baruya, 2015); (1) residues from the harvest and extraction of logs, which are not economically viable for post- processing, but have since become an important feed for pelletizing, (2) residues during timber manufacture, plywood, particleboard, and so on, and (3) waste wood products from transport pallets, construction timber, and furniture. The main types of residue include bark and sawdust. Sawdust is produced as boards and sections are sawn from the log. Accumulated sawdust in sawmills need to be removed quickly due to the fungal growth, which leads to wood rot and can become an obstacle to the production process. Bark, which may have a relatively high moisture content but is still combustible, is often removed from the log before it is sawn into lumber. Bark can be deducted for the production of pellets, however, it is often included in the generation of roundwood, which is produced from the trunk section of trees and contains the source of most of the sawmilling residues, for industrial purposes such as paper production, construction, and furniture production. Wood pellets can be produced by mixing bark and other materials, such as sawdust.

The rate of residue recovery varies between different countries due to various harvesting practices and forest management programs applied, considering different policy schemes and the size of forest area. The Russian Federation, Brazil, Canada, the United States, and China account for more than half of the world's forest area. In the United States, 98% of the bark, sawdust, and wood trimmings from sawmill operations, and the black liquor produced in the pulping process, are currently used as fuel or to produce other fiber products (EIA, 1994). In Asia, on average, only half of the wood harvested is used and the rest is unused residue that goes to waste (Enters, 2001). A reasonable estimate of the residue stock in a given forest mass can be calculated by multiplying the percentage of available residue in a single tree (specified as 46% by Baruya, 2015) by the tones of biomass in a nation's productive forest (Baruya, 2015).

1.1.2.3 Harvesting, Treatment and Storage of Residues

To avoid encountering a management problem that will affect the next season's residue yield in the regions where large quantities of agricultural residue are produced and left on the soil, it is recommended that before residue removal, the impact of this practice on crop production and soil properties must be considered, to maintain the optimum organic matter level of the soil and for agricultural productivity. Management practices such as no-till farming, that is, growing the crops without disturbing the soil through tillage, choice of crops in a rotation and rotation length, and adequate fertilization can be used to enhance levels of organic matter in soil, which might counterbalance the potentially negative impact of residue removal (Iqbal et al., 2016). Straw retention and no tillage showed considerable beneficial influence on these parameters for rapeseed (Malhi et al., 2006). In order to harvest woody biomass, certain functions must be accomplished. Initially, by the felling function, the stems are severed or pulled from the ground. Then, the cut material has to be collected and transported to the roadside (primary transportation function). Somewhere in this process, the biomass material may be converted into a different form like chips, bundles, or bales that is more efficient to handle. Finally, processed biomass is loaded onto a transportation vehicle.

Extraction and transportation of unsorted, uncompacted forest residues are usually uneconomic due to the bulky nature of the loose material, resulting in low payloads and higher haulage cost. Therefore, more productive systems are needed to make the forest residues less bulky and more compact for easier material transfer with higher payloads. Chipping is one these systems, which is commonly used globally to produce wood chips that are less bulky and easier to handle and transport. However, they can be difficult to store, unless dry, as they constitute a reservoir for fungi and also due to self-combustion. To produce and transport chips specialized machinery is required.

There are four main comminution systems utilized for chipping; comminution in the terrain chipping, comminution at a landing (roadside chipping), comminution at a plant, and comminution at a terminal. By comminution in the terrain, forest chips are produced in the terrain, or at source. It requires a highly mobile chipper with a mounted tipping tank for storing chips, which moves in the terrain on strip roads and transfers the biomass with its grapple loader to the feeder of the chipping device. It is difficult to control terrain chipping systems when large volumes of wood chip are produced, hence the role of this system is diminishing in large-scale systems.

Comminution at a landing is performed by either using farm tractor-driven chippers in small-scale operations or heavy truck-mounted chippers in larger operations. Although the large roadside inventories of biomass and the simultaneous presence of the chipper and the truck requires a wider landing area in this system than in the alternative systems, landing chippers are distinguished by the properties of being stronger, more reliable, and more efficient than terrain chippers. When comminution is performed at the plant, the chipper and chip truck are independent of each other and truck transportation of biomass takes place in the form of loose logging residues, whole trees, or pieces of stump and root wood. Hence, the technical and operative availability of the equipment is high, demand for labor is decreased, and control of the procurement process is facilitated. Comminution at a terminal is a compromise between comminution at a landing and at the plant. It facilitates the shorter distance from the logging site to the terminal when the network of terminals is dense.

Bundling is a way of harvesting and handling forest residues, other than chipping, that have the potential to improve the cost competitiveness of residues as an energy source. The bulk density of forest residues is increased so as to obtain more compact, uniform bundles that are easier than loose residues to handle, store, and transport. Hence, handling, storage, and transportation costs are reduced. Bundling also allows the stored material to dry sufficiently to improve its quality as an energy source.

1.1.3 Wastes

1.1.3.1 Animal Wastes

Animal wastes can be used as sources of biomass-based conversion processes, especially in bio-energy and bio-fertilizer production. The most common sources include dairy shed effluent (containing urine, dung, wash water, residual milk, and waste feed), dairy manure, poultry litter (a mix of manure, water, spilled feed, feathers, and bedding material), renderings, and other wastes from livestock finishing operations. Dairy manure and poultry litter are among the most commonly used animal wastes in biomass-based production chains throughout the world. The large volume of manure that a cow produces, and automated operations that lead to frequent manure collection make the dairy industry a well-suited option to biomass-based production. In the past, this large volume of cow manure was treated and sold as fertilizer for agricultural purposes, or simply disposed of through farmland applications, by spreading on agricultural lands. However, the introduction of environmental regulations and legislations that stringently control odor, surface and groundwater pollution, soil contamination, and nutrient management, forces new waste management and disposal practices, which provides further incentives for the use of animal manure in biomass-based production chains.

The phosphorus (P) run-off from concentrated poultry areas to surface water creates ecological issues such as odor and taste problems in drinking water, due to excess algae growth because of P enrichment of the water. Local land applications of poultry litter are restricted because of these water quality concerns, and alternatives to local land application will be needed for much of the poultry litter produced in concentrated poultry areas (Bock, 1999). Due to its high methane content, poultry litter is an efficient feedstock for biomass to bio-fuel and bio-energy conversion processes such as anaerobic digestion. It is suggested that the methane yield from the anaerobic digestion of poultry waste is higher than that from the digestion of piggery and cattle waste (Itodo and Awulu, 1999). In addition, the poultry litter contains a higher fraction of biodegradable organic matter than other livestock wastes and this includes high levels of organic nitrogen due to the high content of protein and amino acids, hence the digestate obtained from the anaerobic digestion process can be used as high-quality organic fertilizer for agricultural purposes (Itodo and Awulu, 1999). Anaerobic digestion is one of the most common ways of converting dairy manure and poultry litter into useful forms, by which biogas, bioenergy, and nutrient-rich digestate (used as a bio-fertilizer) can be produced simultaneously. Combustion, pyrolysis, gasification, and composting are other processes for converting animal wastes into various bio-products. These conversion processes can be referred to and are described in more detail in Chapter 2 of this book. The most appropriate conversion process can be chosen considering the amount and type of the waste produced by dairy or poultry operations, the nutrient and energy content of the waste, moisture content of the waste, and the amount of contaminants.

1.1.3.2 Municipal Solid Waste

The US Environmental Protection Agency defined Municipal Solid Waste (MSW) as any "solid phase household, commercial/retail, and/or institutional waste." In other words, MSW encompasses nonhazardous mixed trash of everyday items collected from residential areas, commercial buildings (offices, retail and wholesale establishments, restaurants), and institutions (libraries, schools, hospitals), such as food waste, yard waste, papers, plastic bags, bottles, metals, parts of furniture, batteries, and clothing. On the contrary to more homogeneous waste streams resulting from agricultural and industrial processes, MSW comprises numerous types of waste, which makes it the most complex waste stream. Various European countries, such as Germany, the United Kingdom, and France, as well as numerous states in the US, have adopted regulations and legislations for disposal and management of MSW, to promote recycling and reduce the waste destined for landfill, which is becoming environmentally and economically less desirable, due to the toxic decomposition of the material in landfill areas, wastage of valuable areas, and increasing transportation cost. Therefore, various technologies and processes have been developed to divert MSW, primarily disposed to landfills, which have the potential to handle and manage the MSW in a more sustainable way than via landfill. These include waste-to-energy technologies, material recovery operations through recycling, and composting of wastes with a high organic content.

Waste-to-energy technologies can play a vital role in mitigating the problems by presenting the opportunity for utilizing MSW to produce energy in terms of electricity and thermal energy, besides an alternative to landfilling or composting the MSW. A significant reduction in the overall waste quantities requiring final disposal can be achieved by these technologies. By mass burn incineration, one of the most commonly applied techniques to convert waste into energy, unprocessed or minimally processed MSW can be completely combusted in waste-to-energy facilities as a fuel. At these facilities, heat generated from the combustion process can be distributed by district heating networks to be utilized in residential, industrial, or agricultural operations, or in power turbine generators for electricity production, after turning water into steam by the generated heat. The waste-to-energy facility can be paid by the waste suppliers to take the fuel, different from many other energy technologies, where fuel is required to be purchased. This fee is known as "tipping fee," of which amount depends on the agreement between fuel suppliers and the waste-to-energy facility.

Some technical, economical, and environmental disadvantages of conventional incineration processes lead to growing attention to other thermal processes for the treatment of solid wastes. One of the main reasons of dissatisfaction from incineration processes is that the combustion products are highly erosive and contain high levels of chloride that lead to high temperature corrosion in heat exchange tube surfaces. Gasification and pyrolysis are considered as emerging and alternative thermal processes to produce energy from MSW, and which happen under different amounts of air presence in the system. Thermal decomposition of organic material, producing a mixture of combustible gases, containing methane as well as liquids and solid residues, takes place in pyrolysis at high temperatures in the absence of air, whereas the thermal decomposition is realized in the presence of a limited amount of air by gasification of MSW. Then combustible gas produced by both processes can then be used in either boilers or in combustion turbines after being cleaned up to produce electrical and/or thermal energy. Although both pyrolysis and gasification are proved to be efficient in processing biomass and homogeneous industrial wastes, their application on MSW conversion is still under development. Among them, gasification has been used in different applications with various feedstock alternatives and shows potential for processing MSW. Another MSW-to-energy technology is anaerobic digestion, by which electrical and thermal energy is generated from the landfill gas produced through the biological degradation of the organic fraction of MSW by anaerobic fermentation (also referred as digestion). The generated gas, containing a high level of methane and structurally similar to natural gas can be either fed into the national gas grid to be used for domestic, industrial, or agricultural purposes, or converted into electricity and heat in energy generation turbines.

1.1.3.3 Wastewater and Sewage Sludge

The term "wastewater" is basically used for "waste of all types of used water from residential, industrial, business, and agricultural areas." The area that water is obtained from also specifies the type of wastewater, that is, agricultural wastewater, municipal wastewater, industrial wastewater. The main types of wastewater used as feedstock in biomass-based production systems include domestic greywater (wastewater from baths and kitchens), blackwater (wastewater from toilets), residual water from industrial processes, residual water from commercial or municipal institutions, rainwater, and residual water from agricultural and husbandry processes. A mixture of these types can be utilized as well. Besides procurement in biomass-based production plants, wastewater can be used for irrigation, aquaculture, forestry, and industrial activities, generally after some treatment processes. The application area is changed according to the level and type of treatment process.

Sewage sludge is the specific type of wastewater obtained from municipal pipeline network (also known as sewerage). Excreta, also called fecal sludge, contains sludge, scum, and liquid, which can be collected on-site by pumping it into the trucks that are used for transportation to treatment ponds or to the discharge location for untreated material. In sewage treatment facilities, sewage sludge is produced by removing solids from wastewater and converting the inherent organic compounds into a bacterial biomass. Although untreated sludge has many disadvantages such as having many harmful pollutants and pathogens, as well as putridity, it includes many beneficial nutrients for plants and can be used a very efficient fertilizer and soil conditioner, which increases the physical and chemical fertility of soil, once treated. The basic properties of sludge, which specify the quality of treated material as well, are highly related to the treatment level and type (see Section 1.3.5), its origin (residential area or public places), and quantity of flushing water contained in it. Sewage sludge is commonly used for the generation of energy and energy carriers either directly after a drying process by incineration, or indirectly via gasification, pyrolysis, or anaerobic digestion systems. Anaerobic digestion is especially suitable for the sludge that does not contain heavy metals, whereas pyrolysis, gasification, or incineration may be appropriate when there are persistent impurities in sludge. Composting is a bacterial degradation process occurring in the presence of oxygen, which can be used either as a treatment method to stabilize the organic content in the sludge prior to the energy recovery process or to convert the sludge into a compost that can be used as a high-quality soil conditioner with high nutrient and organic carbon content.

1.1.3.4 Industrial Wastes

Many industries and commercial operations produce a wide range of liquid and solid wastes with organic content that can be used as feedstock in biomass-based production systems. These operations include food and beverage processing, production of pharmaceuticals and organic chemicals, tannery, and pulp and paper processing. The wastes from these operations can be utilized as efficient substrates in anaerobic digestion systems to produce biomethane and bio-fertilizer. Biochemical degradation of some of these wastes might be difficult when used as a sole substrate, however, they can easily be digested in combination with other biomass sources with organic content. Considering this fact, co-digestion of industrial wastes with animal waste or sewage sludge have gained considerable attention and popularity worldwide in recent decades.

Industries based on food and beverage processing, such as meat and dairy processing and preserving, vegetable oil production, fruit and vegetable canning and preserving, production of soft drinks and milk, and production of sugar and sugar confectionery, generate various kind of organic wastes and by-products to be used as biomass feedstock. Liquid wastes from these industries include, water from washing, cleaning, precooking and blanching poultry, fish, meat, fruit, and vegetables, whereas solid wastes originate from peeling and scraping fruit and vegetables, waste food that is rotten and does not meet certain quality standards, coffee grounds, pulp and fiber from the extraction of sugar and starch.

Tannery operations, which comprise all phases of leather-making, produce a large amount of highly polluting waste water involving compounds of leather cuttings, trimmings and fleshing residue that is often disposed in natural water bodies, especially in developing countries. Another industrial process that generates contaminated organic waste and consumes a significant amount of water and energy is paper and pulp production. Pulp and paper industries discharge a highly heterogeneous wastewater containing pieces of wood and chemicals. Disposal of the wastewater from the tannery and pulp and paper industries results in some major problems for the stakeholders including obnoxious odors, contamination of the water used for agricultural and domestic purposes such as bathing and cooking, contamination of the soil and groundwater system that provide drinking water, and in endangering the lives of aquatic animals. Besides producing bio-fuels, bio-energy, bio-chemicals, and other types of useful bio-products, making use of the wastewater from the tannery and pulp and paper industries in biomass-based production chains prevents the disposal of a large amount of contaminated wastewater into the soil and water bodies.

1.1.3.5 Treatment and Storage of Wastes

Its varying composition and size as well as undesirable materials contained may ruin the process or emission control systems, therefore preprocessing operations are necessary, which convert MSW into a more suitable feed material for thermal processes, to elevate the performance of system. Preprocessing of MSW combines manual and mechanical activities such as material separation and sorting, grinding and shredding, blending and mixing with other materials, drying, baling, and pelletization. The primary aim is producing, a feed material with consistent physical characteristics and chemical properties that can be handled, transported, and stored efficiently and safely.

MSW can also be preprocessed and treated to separate out recyclable and/or noncombustible components such as ferrous materials, glass, stones and grit, and other materials, and be converted into refuse-derived fuel (RDF). Hence, RDF mainly comprises combustible elements of MSW, such as nonrecyclable plastics, paper, and labels. Incoming MSW can be exposed to different processing steps, such as magnetic separation of ferrous and nonferrous materials, mechanical screening to separate out glass, air classification to separate the light materials from the heavy ones and shredding into a uniform grain size using mechanical methods, using RDF systems to produce a more homogenous and combustible mixture that can be used as improved fuel in combustion or boiler systems. Typically, the light materials which also have an elevated calorific value create the final RDF whereas the heavy ones are destined for landfills. The final product may be either dried and compacted into pellets or sold in its processed form as a plain mixture.

Initial characteristics and required final use of sludge, which is a concentrated stream of solids in liquid, specifies the treatment method. Since sludge typically has a high water content, harmful pollutants and pathogens, putridity pretreatment should be applied prior to the final conversion process, to generate energy and/or soil amendment. Dewatering (drying), thickening and stabilization (composting) are the most commonly used pretreatment techniques applied for sewage sludge. Thickening reduces the total volume of sludge using a gravity thickener, since it is more practical in handling thicker sludge. Dissolved-air flotation is another method for thickening, in which suspended solids in sludge are carried to the surface by air bubbles to form a layer of thickened sludge. By dewatering (drying), the amount of water is decreased by combining gravity drainage and evaporation techniques. Although there is significant amount of water still in the dried material, the sludge is not in liquid form any more, hence the water content does not hinder the handling of the sludge as a solid material. Stabilization is applied for lessening the biochemical oxygen demand of sludge, which makes it extremely difficult to dewater. Stabilization can be done as a part of the composting process, which requires an increased solid content to be handled easily in solid form. Wastewater treatment plants have two major aims, reducing the volume of sludge to reduce the storage and pumping costs and stabilizing the organic materials in it to minimize the obnoxious odor, before final disposal.

To enhance the wastewater reuse especially for agricultural purposes, raw municipal wastewater is exposed to preliminary, primary, secondary, and in some cases, further treatment processes to obtain sludge in some countries, such as the United States. Preliminary wastewater treatment basically includes screening (by separating coarse solids that may break down the equipment) and grit removal (by removing inorganic, heavy solids that may hinder the treatment process), prior to the primary treatment, by which settleable matter is removed by gravity sedimentation. A biological process is applied in the secondary treatment of wastewater, such as activated sludge or digestion for removing biodegradable organic material, that is, nutrients. The level of treatment greatly depends on the type of agricultural activity (i.e., type of the crop), and local geographic and soil conditions.

As minimizing the level of degradation of the organic material in waste biomass sources is important to obtain high yields of bio-products by conversion processes, these biomass sources should be kept in storage for adequate time periods. Long-term waste storage may lead to considerable levels of degradation in the organic content of waste biomass. However, in practical cases a minimum period might have been set by the company for keeping the biomass feedstock in storage due to inventory management and production planning strategies. Hence, an optimized and adequate duration and timing should be specified for the storage of waste-type biomass sources, to both minimize the level of degradation of the organic material for obtaining high yields of bio-products by conversion processes and to align with the inventory and production policies.

Prior to the conversion into various kinds of bio-products in biomass-based supply chains, waste biomass is often stored in containers of an approved size, type, and construction. Containers for the storage of waste biomass should be suitable for the climatic conditions of the area, watertight, insect and rodent resistant, and not easily corrodible. High volume containers are often needed to store the sewage sludge and wastewater. Chemical composition (e.g., solids content, percentage of volatile material) and granularity of sludge (e.g., dust and sand presence, particle size), are essential factors in determining the design, capacity, and dimensions of sludge storage containers, of which type and characteristics are important to ensure an efficient treatment process, enhance availability, and safety.

Facilities for sludge storage may involve lagoons, drying beds and separate tanks to store liquid and dewatered or dried sludge. For the stabilization of bio-solids, additional volume may be required, however, due to space limitations in the treatment and storage area this is generally limited in capacity. Completely stabilized sludge may be stored in lagoons. Lagoons should be placed as far as practicable from residential areas and handled so as to prevent leaching into the groundwater or adjacent soils to prevent environmental contamination. Lagoons can be constructed either permanently near the sewage treatment plant or temporarily for bio-solid utilization on agricultural land. Sludge with a high pH (usually >12) should be stored in a tank or vessel that enables rapid disposal or re-treatment of the sludge in case of pH reduction, instead of a lagoon. In most practical applications, on-site storage of dewatered alkaline stabilized sludge changes between one week and three months depending on the solids content of the sludge. Dried sludge may be stored on-site in bins or covered facilities by taking necessary precautions for explosive gaseous by-products, odors, or dust.

Solid wastes can be kept in bulk storage areas, of which size depends on the quantity and density of the bio-solids, their moisture content and when it will be used in conversion or pretreatment systems. Solids with lower moisture content can be stacked in storage areas with less slumping possibilities. Additional areas should be provided for cleaning and maintenance operations. Odor control should be provided to reduce the impact on the surrounding residential areas especially in the bulk storage areas with destabilized bio-solids.

1.2 RECENT TRENDS IN BIOMASS PRODUCTION

In recent decades, research efforts have been made to explore new biomass sources for improving biomass-based production yield, mainly focusing on bio-fuel production. Among these emerging sources, aquatic species, especially algae, attract considerable attention of researchers. One of the most comprehensive studies in this field is the US National Renewable Energy Laboratory's experiment, using algae as a bio-diesel source in the scope of the "Aquatic Species Program" from 1978 to 1996 (Sheehan et al., 1998). Many algae species such as microalgae, macroalgae, and cyanobacteria perform photosynthesis and make efficient use of light and nutrients to enhance faster biomass growth rates. An advantage of algae production is that they have a short cycle of harvesting (typically 1–10 days) that allows several harvests in a short time period contrary to the annual energy crops. In addition, microalgae can easily be cultivated in an aquatic environment on land unsuitable for the cultivation of other crops, such as land with an excessively saline soil, arid, or drought-stricken land, which minimizes the competition on arable land with the cultivation of food crops. Various types of fuel can be produced by using algae as feedstock. For example, algae, which are oil-rich plants with a natural oil content greater than 50%, can be produced at wastewater treatment plants using algae ponds, and then be extracted from the systems to be converted into bio-diesel (Briggs, 2004). Furthermore, bioethanol can be generated by fermentation of sugars in algae, as a separate process or just after the extraction of its oily part. Algae is an extremely efficient and high-yield biomass source that grows faster than conventional crops, and can be converted into hundreds of times more oil per unit of area than conventional crops such as rapeseed, soybeans, or jatropha. Besides its projected high yield, one of the advantages of algae as an energy source is that, it does not entail a decrease in food production, since it requires neither farmland nor fresh water. Not only research efforts but also commercializing studies are ongoing for algae production for energy purposes.

Another emerging resource for biomass-based production is jatropha, which is an oily poisonous tree. The seeds of jatropha tree, which have an oil content of approximately 40%, can be considered as an efficient feedstock for mechanical extraction and bio-diesel production. The oil obtained from jatropha has been a valuable fuel that substitutes bio-diesel by being directly used, after extraction, in diesel engines. Multidisciplinary research is being conducted on improving the oil yield of jatropha, integrating biotechnology, agronomy, genetics, and soil science disciplines. Research efforts have been made to explore the viability of bred mustard varieties as a source for bio-fuel production. It has been shown that using bred mustard leads to a high oil yield and the residual cake occurring after the oil is extracted can be used as an effective biodegradable pesticide. Other emerging biomass sources, of which the use in generating bio-fuel has been investigated and discussed, include fungal species, pongam oiltree, and used coffee grounds.

1.3 DISCUSSION

Interest in renewable and sustainable resources have increased in parallel with the realization that nonrenewable-based production systems induce increase in greenhouse gas emissions to the atmosphere as well as other environmental damages, draw on finite and expensive sources that have limited existence in the environment, and that they are indeed not economical, as most of the countries have to import all these type of resources from different parts of the world. The disadvantages of using nonrenewable resources make it essential to explore new and alternative sources that are naturally replenished and will never run out, hence ensuring the sustainability of supply and production with minimum negative impact on the environment. Due to these facts, biomass, which is a biodegradable material that is converted into end products through various conversion processes, has become a significant alternative source of renewable-based production in recent years. To effectively design the configuration of biomass-based production chains and manage the operational activities, decision-maker(s) should have a detailed knowledge about available biomass resources as well as the basics about production, collection, handling, pretreatment, and storage of these resources that are or can be utilized throughout the chain. Considering this fact, this chapter aims at providing a comprehensive understanding on various biomass sources that are commonly used as feedstock in biomass supply chains, focusing on the information to be used in decision-making procedures for the design and management of biomass-based production chains.

Dedicated energy crops are efficient biomass sources, from which high yields can be obtained with low input, and they reduce soil erosion and improve soil quality. In addition, they can provide up to five or six times more potential energy than the amount of energy required to grow and produce them. To obtain the potentially available yields from energy crop cultivation, agricultural practices need to be changed, however, important climate and water quality benefits can be especially obtained by applications of perennial grasses. A number of uncertainties, in geographic (e.g., soil and water conditions), climatic (i.e., weather conditions), natural and operational conditions of the cultivation area, affect the yield of dedicated energy crops (for detailed information on these uncertainties refer to Chapter 5). Other factors that have a significant effect on the yields are; species, cultivar, and planting density, agricultural procedures and practices, legislative measures and standards, and harvest cycles. Moisture content is a major parameter determining the energy content of the crop. Hence, it is critical to use proper design and modeling approaches, prior to cultivation of energy crops in dedicated land, to optimize locational and operational factors and to increase the productivity by eliminating the negative effects of inherent uncertainties on the yields.

Utilizing wastes and residues in biomass-based production chains as feedstock, not only leads to generate bio-products in a resource efficient way, but also mitigates an environmental management issue by presenting attractive options for many of the challenges for disposal of these materials. A slight change in consumption trends because of the development of urbanization and change in income pattern may cause substantial differences in quantum and diversity of waste and residue materials as well as their harmful effects on the environment and public health. This fact poses a greater challenge and urgency for the municipalities to handle the large quantities of waste and residue with proper waste management programs which result in an increasing awareness about an urgent need to adopt scientific methods to safely dispose of these materials. To cope with the issues, owing to the necessity of handling and/or disposing of large waste/residue streams, proper methodologies have to be developed and applied to optimize operational activities considering proper waste management strategies as well as the main characteristics of these wastes/residues.

REFERENCES

Adams, P.W.R., Lindegaard, K., 2016. A critical appraisal of the effectiveness of UK perennial energy crops policy since 1990. Renew. Sust. Energ. Rev. 55, 188–202.

Allen, B., Kretschmer, B., Baldock, D. et al., 2014. Space for energy crops—assessing the potential contribution to Europe's energy future. Report produced for BirdLife Europe, European Environmental Bureau and Transport & Environment. IEEP, London.

Baruya, P., 2015. World forest and agricultural crop residue resources for cofiring. IEA Clean Coal Centre.

Bock, B.R., 1999. Poultry litter to energy: technical and economic feasibility. <http://www.brbock.com/RefFiles/PoultryLitter_Energy.doc>.

Briggs, M., 2004. Widescale bio-diesel production from algae. University of New Hampshire, Durham, NH.

Caslin, B., Finnan, J., Mccracken, A., 2010. Short-rotation coppice willow best practice guidelines. Agri-Food and Biosciences Institute, Belfast.

DEFRA, 2007. Planting and growing miscanthus, Best practice guidelines for Applicants to Defra's Energy Crops Scheme. Department for Environment, Food and Rural Affairs, London. <http://www.naturalengland.org.uk/planning/grants-funding/energy-crops/default.htm>.

Don, A., Osborne, B., Hastings, A., et al., 2012. Land-use change to bio-energy production in Europe: implications for the greenhouse gas balance and soil carbon. GCB Bioenergy 4, 372–391.

EIA, 1994. Manufacturing consumption of energy, US Department of Energy report DOE/EIA-0512(91). US Government Printing Office, Washington, DC.

Enters, T., 2001. Trash or treasure? Logging and mill residues in Asia and the Pacific, RAP publication 2001/16. FAO, Bangkok, Thailand.

Fernando, A.L., Costa, J., Barbosa, B., et al., 2017. Environmental impact assessment of perennial crops cultivation on marginal soils in the Mediterranean Region. Biomass Bioenergy . Available from: https://doi.org/10.1016/j.biombioe.2017.04.005.

Hauk, S., Gandorfer, M., Wittkopf, S., et al., 2017. Ecological diversification is risk reducing and economically profitable—the case of biomass production with short rotation woody crops in south German land-use portfolios. Biomass Bioenergy 98, 142–152.

Hinchee, M., Rottmann, W., Mullinax, L., et al., 2009. Short-rotation woody crops for bioenergy and bio-fuels applications. In Vitro Cell. Dev. Biol. Plant 45 (6), 619–629.

Hoogwijk, M.M., Faaij, A., van den Broek, R., et al., 2003. Exploration of the ranges of the global potential of biomass for energy. Biomass Bioenergy 25 (2), 119–133.

Iqbal, Y., Lewandowski, I., Weinreich, A., et al., 2016. Maximizing the yield of biomass from residues of agricultural crops and biomass from forestry, Final report. European Commission, DGEnergy.

Itodo, I.N., Awulu, J.O., 1999. Effects of total solids concentrations of poultry, cattle, and piggery waste slurries on biogas yield. Trans. ASAE. 42 (6), 1853–1855.

Kretschmer, U., Kirchner, N., Morhart, C., Spiecker, H., 2013. A new approach to assessing tree stem quality characteristics using terrestrial laser scans. Silva Fennica 47 (5), article id 1071.

Lal, R., 2004. Soil carbon sequestration impacts on global climate change and food security. Science 304 (5677), 1623–1627.

Malhi, Y., Wood, D., Baker, T.R., et al., 2006. The regional variation of aboveground live biomass in old-growth Amazonian forests. Glob. Chang. Biol. 12 (7), 1107–1138.

Miao, Z., Shastri, Y., Grift, T.E., et al., 2012. Lignocellulosic biomass feedstock transportation alternatives, logistics, equipment configurations, and modeling. Biofuel, Bioprod. Bior. 6, 351–362.

Mola-Yudego, B., González-Olabarria, J.R., 2010. Mapping the expansion and distribution of willow plantations for bio-energy in Sweden: lessons to be learned about the spread of energy crops. Biomass Bioenergy 34 (4), 442–448.

Monforti, F., Bodis, K., Scarlat, N., Dallemand, J.-F., 2013. The possible contribution of agricultural crop residues to renewable energy targets in Europe: a spatially explicit study. Renew. Sust. Energ. Rev 19, 666–677.

Ocumpaugh, W.R., Kuntz, D., Rahmes, J. et al., 2003. Desmanthus bicornutus: a new summer-growing perennial shrubby legume for south Texas and Mexico. Proceedings of the American Forage and Grassland Council, vol. 12.

Panoutsou, C., Eleftheriadis, J., Nikolaou, A., 2009. Biomass supply in EU27 from 2010 to 2030. Energy Policy 37 (12), 5675–5686.

Sheehan, J., Dunahay, T., Benemann, J., Roessler, P., 1998. A look back at the US Department of Energy's Aquatic Species Program—bio-diesel from Algae. US Department of Energy's Office of Fuels Development.

Stewart, C.E., Follett, R.F., Pruessner, E.G., et al., 2015. Nitrogen and harvest effects on soil properties underrainfed switchgrass and no-till corn over 9 years: implications for soil quality. GCB Bioenergy 7, 288–301.

Turhollow, A., Wilkerson, E., Sokhansanj, S., 2009. Cost methodology for biomass feedstocks: herbaceous crops and agricultural residues. Oak Ridge National Laboratory.

Wolf, D.D., Fiske, D.A., 1995. Planting and managing switchgrass for forage wildlife, and conservation. Virginia Cooperative Extension Pub. 418–013. Virginia Cooperative Extension, Blacksburg, VA.

Wright, L.L., Perlack, R.D., Turhollow, A.F., Eaton, L.M., 2011. Switchgrass production in the USA, IEA Bioenergy Task 43.

FURTHER READING

Hongtao, W., Yongfeng, N., 2001. Municipal solid waste characteristics and management in China. J. Air Waste Manage Assoc. 51 (2), 250–263. Available from: https://doi.org/10.1080/10473289.2001.10464266.

Chapter 2

Biomass-Based Production Systems

Chapter Outline

2.1 The Main Systems, Technologies, and Processes for Biomass-Based Production 25
2.1.1 Thermochemical Conversion Technologies 26
2.1.2 Biochemical Conversion Technologies 41
2.1.3 Biomass Extraction 49
2.2 Recent Trends in Biomass-Based Production Technologies 51
2.3 Discussion 53
References 54

2.1 THE MAIN SYSTEMS, TECHNOLOGIES, AND PROCESSES FOR BIOMASS-BASED PRODUCTION

Biomass sources have a great potential as renewable feedstock to be converted into useful forms of energy (i.e., electrical energy, thermal energy, and energy carriers/bio-fuels), with a wide range of conversion technology and process options with different scales. However, in order to provide an economically and environmentally efficient process to compete with fossil fuel-based energy production systems, proper conversion technologies need to be selected. Among all the technology options given for defined system components and power generation, not all combinations may be sensible from an efficiency, environmental, and economic point of view. Hence, selecting the appropriate process is important in terms of obtaining optimal design of plant configurations and management of operations. Type and quantity of biomass feedstock used by conversion systems, type of biomass pretreatment technology, desired form of the energy according to end use requirements, environmental standards and conditions, and economic conditions and initiatives are the main factors that influence the choice of conversion process.

Technologies for conversion of biomass into bio-products can be categorized into three main classes: thermochemical processes, bio-chemical/biological processes, and extraction. Fig. 2.1 depicts the main conversion routes

Decision-Making for Biomass-Based Production Chains. DOI: https://doi.org/10.1016/B978-0-12-814278-3.00002-9

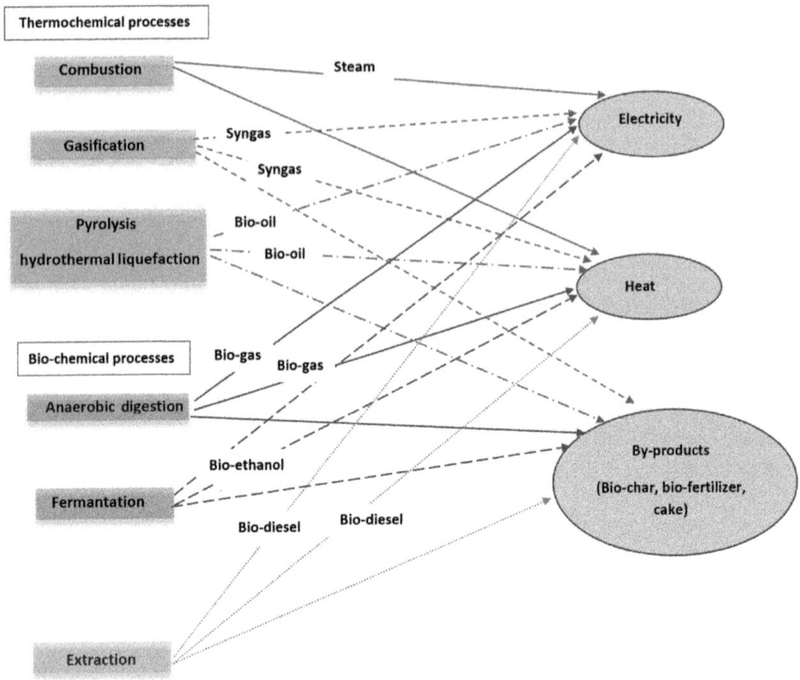

FIGURE 2.1 The main conversion routes for biomass.

for biomass. The conversion processes in these three categories are described in the following subsections.

2.1.1 Thermochemical Conversion Technologies

Thermochemical conversion technologies apply heat and chemical processes to produce energy and chemicals from biomass. Within thermochemical conversion, four commonly used process options are explained in this section; combustion, gasification, pyrolysis, and hydrothermal liquefaction (HTL).

2.1.1.1 Combustion

Combustion is a thermochemical reaction in which biomass fuel is burnt, generally in the presence of atmospheric air or sometimes only oxygen. For millennia, direct combustion of biomass has been a basic and well-known technology used for converting biomass into heat and, later on, for generating power through steam.

Feedstock: The most commonly used biomass feedstock for combustion is wood and lignocellulosic material, however, a wide range of industrial and processing residues that have different material properties and combustion characteristics can be effectively utilized as feedstock in this process. These include residuals and by-products that have a high total solid content such as bark residuals, straw, sawdust, and shavings from sawmills. Energy crops that are specifically produced to create feedstock for energy generation, such as switchgrass, poplar, and willow are also efficient materials for combustion. Because of some characteristics of biomass resources, such as low energy density, they are more difficult to handle, transport, and combust than fossil fuels. Pretreatment technologies, such as pelletizing, are increasingly popular to make the handling of agricultural and wood residues easier. Another problem that lead to difficulties in using biomass as combustion fuel is related to its inorganic content. The release of these inorganic components, such as salt, chlorine, sulfur, and potassium, in the combustion process may cause reduced heat transfer as a result of heavy deposition on heat transfer surfaces that may cause operational disruptions and hence unscheduled shutdowns.

An important characteristic of biomass feedstock is the high level of volatile material that is released during the combustion process. Maximizing combustion of this volatile material is very important for both a high combustion quality and a low emission level. Temperature of the combustion chamber, residence time, the amount of air, and the turbulence of the burning gases are the most important parameters that require attention, to obtain an efficient and environmentally satisfactory combustion process. Combustion technology and settings (i.e., design and size of combustion chamber, process control technology, primary and secondary air ratio, distribution of the air nozzles), fuel characteristics (i.e., shape and size of fuel, moisture content, organic content, ash content, ash melting behavior) and load condition (full- or part-load) are the technical factors that have the most significant effect on these parameters. In addition, a good contact should be provided between the oxygen in the air and the biomass fuel, to increase the efficiency of the combustion process.

Process description: Biomass combustion is the oldest and the most common form of combustion used by humanity, however, it is one of the most complex biomass-based production systems to manage, due to the use of solid fuels in a multistage reaction system, which requires extensive interaction between the thermal and mass fluxes. Hence, it is critical to properly analyze the system conditions and requirements and design efficient combustion systems. Carbon dioxide and water vapor are the main outputs of the combustion process, however, considering the significant amounts of smoke, tar, and ash particles emitted, the design and optimization of environmentally acceptable biomass combustion systems are important in terms of the minimization of these emissions and the accommodation of their possible harmful effects. In the first phase of combustion, namely the pyrolysis phase, combustible vapors from the biomass burn as flames, which is also called fuel gas evolution. The pyrolysis stage converts the biomass respectively into gases, volatile liquids, and a solid residue called char. The rate of fuel gas evolution in this stage is affected by both the temperature and the intensity of the heat flux supplied to the solid surface. After all of the volatiles have been removed, the char combustion stage is reached, in which the action of the hot char on water vapor and carbon dioxide is to convert the fully oxidized carbon and hydrogen forms back to fuel intermediates. The energy required for this process comes from the flame in which the carbon monoxide and hydrogen burn with oxygen from the air producing water vapor and carbon dioxide.

Because of a series of limitations, the combustion stages, which are explained above, the overall efficiency of heat generation is limited as well as the maximum temperature reached in the process. As mentioned previously, time, temperature, and turbulence are important parameters for an efficient combustion. These parameters should be maximized simultaneously to maximize the yield from the process. Turbulence is a significant factor for ensuring complete burning, by providing a full and complete mix of the fuel gases with oxygen. Time is essential as the drying and pyrolyzing stages require some time, as well as the combustion of the fuel gases. Eventually, maximized temperature satisfactorily increases the rate of all of the reactions. Using the heat from combustion of biomass results in some practical difficulties on maximizing time, turbulence, and temperature. For example, combustion systems with the highest efficiency have a separate area for combustion from the surfaces for heat transfer, so that in an optimized system, combustion should be fully completed before transferring the heat to the boiler, stove, or any other application area. On the contrary, the systems with a low efficiency, that is, open fires, lose energy by radiation into the space, and by convection of a significant amount of excess air, and result in low process temperatures that cause large unburnt carbon losses.

Technologies: Various equipment such as stoves, boilers, furnaces, steam turbines, and turbo-generators are used to convert the chemical energy in biomass into heat or power by combustion. One of the recent technological advances for biomass combustion, pellet stoves, can operate for at least 24 hours without being tended by delivering a steady supply of compressed pellets of wood or other biomass into the fire. For relatively larger scale applications, biomass-fed boilers can be used to increase the temperature of a building or greenhouse, to meet hot water needs, or to supply steam to power equipment. At the system level, a number of designs of combustion furnaces are available around the world to deal with the different characteristics of various fuel types and combustion requirements. The main technologies can be classified into two categories: fixed-bed combustion systems and fluidized-bed combustion systems. Underfeed stokers and grate firings are two prominent types of fixed-bed combustion systems, in which combustion air is mainly supplied through the grate where the initial combustion of solid fuel takes place and some gasification occurs. Fuels that have high moisture and ash content with varying sizes can be well accommodated by grate firings, however, underfeed stokers are relatively cheap and safe for biomass combustion and they are generally suitable for only small-scale applications. In addition, in grate-fired systems, corrosion, and deposition problems may be the major concern.

Fluidized-Bed Combustion Systems, which are frequently used for wood biomass and waste material, can be investigated in two main types; Bubbling Fluidized Bed and Circulating Fluidized Bed. Bubbling Fluidized Beds not only cope with fuel with varying particle size and moisture content, but also mixtures of different fuel types (i.e., wood and straw) can be combusted by these systems. Circulating Fluidized Bed Furnaces are good at providing very stable conditions for the combustion process, however, the users may encounter some problems related to fuel size, which must be very small, and also there may be some difficulties in running them at partial load. In addition, their larger size makes their cost relatively higher compared to other combustion methods and make them a preferred choice in larger units. One major disadvantage of fluidized bed combustion systems is that alkali metals in the biomass may agglomerate on the bed material, which makes using this technology for biomass, which is rich in alkali metals (e.g., herbaceous-based biomass), disadvantageous. Due to the specific advantages and disadvantages of different combustion technologies, choice of the most suitable technology has a significant effect on the nature and severity of the possible operational problems that can be encountered in biomass combustion.

2.1.1.2 Gasification

Gasification is the conversion of biomass, through partial oxidation, into a gaseous mixture, small quantities of char, and condensable compounds. It is considered as one of the most efficient ways of converting the energy embedded in biomass, and it is becoming one of the best alternatives for the reuse of waste solids (Puig-Arnavat et al., 2010). The fuel gas produced by gasification can be further converted or cleaned prior to combustion. Generation of a combustible gas mixture from solid fuel such as coal or biomass, namely producer gas, by a partial combustion route with restricted air supply (less than required for full combustion) is called gasification.

Feedstock: Biomass suitable for gasification comprises a wide range of materials, such as energy crops, agricultural residues, wood pellets, construction and demolition wastes, lumbering and timbering wastes, and bio-solids (i.e., treated sewage sludge). Woody biomass can be used in the form of dense briquettes made from loose biomass or loose, pulverized biomass.

For efficient implementation, most gasifier systems require uniform-sized biomass fuel. Therefore, biomass that is supplied in a range of sizes, is pretreated to obtain a uniform size or shape. This enables the feeding of the biomass into the gasifier at a consistent rate to ensure that the maximum amount of biomass is gasified for process efficiency. The effectiveness of a gasification process is highly relevant to the characteristics of biomass fuel, such as, moisture content, dust content, tar content, energy content, bulk density, ash, and slagging characteristics. *Moisture content;* the temperature inside the gasifier, hence the efficiency of the gasifier, depends on the moisture content of the biomass fuel. There may be considerable heat loss due to its evaporation prior to gasification. To obtain an optimized temperature level in the reactor, in many gasification systems the biomass fuel needs to be dried to reduce the moisture content before being fed into the gasifier. *Dust content;* the dust that is produced by biomass fuel may be a problematic as it can clog the internal combustion engine, furthermore, it may cause more load on the filters. These issues necessitate frequent flushing and increased maintenance that increase the cost of the system. *Tar content;* Tar is the product of a highly irreversible process taking place in the pyrolysis zone. It tends to deposit in the carburetor that may cause unscheduled shutdowns. Hence, it is one of the most undesirable contents of the producer gas. However, most of the research in this field has focused on cleaning this tar by filters and coolers, but very little effort has been devoted in the area of removing or burning tar in the gasifier to obtain a relatively tar-free gas. As well as dust and tar content, the *content and composition of ash*, which is basically the mineral content in the fuel that remains after complete combustion, of biomass fuel has a major impact on the operation of gasifier. The ash converts into slag which may inhibit or stop the flow of biomass feed in its normal, downward path. Slagging can be coped with by keeping the process temperature either below the flow temperature or above the melting point of the ash. The most appropriate way can be chosen based on specific fuel and gasifier design.

Process description: The reactor for this process is called a gasifier. A gasifier mainly consists of a reactor where the combustible gas is produced, and a cooling and cleaning unit follows the reaction phase to cool and clean the gas ready for power generation in gas engines. The following four main stages occur inside a biomass gasifier during the gasification process. In each phase, different chemical and thermal reactions take place, therefore can be assumed to occupy a separate zone in the gasifier, although there is a considerable overlap of these consecutive stages.

Drying: In the drying stage, biomass fuel, usually with 10%–35% moisture content, is heated to about 100 °C, to convert the moisture into steam.

Pyrolysis: The biomass fuel undergoes the pyrolysis stage after drying. In this stage, biomass burns completely without supplying oxygen, resulting in decomposition of the biomass into solid, liquid, and gaseous parts, which are charcoal, tar, and flue gases, respectively.

Oxidation: After the decomposition of biomass, air is introduced into the gasifier and oxidation takes place. In this phase, carbon dioxide and heat are produced through the reaction of charcoal with the oxygen in the air.

Reduction: In the reduction phase, carbon monoxide (producer gas) is produced through absorption of heat by carbon dioxide and reaction with the remaining char from the pyrolysis stage. Hydrogen is produced as a secondary fuel component by the water vapor in the gasifier. Hence a synthesis gas, namely syngas, is generated that basically comprises hydrogen and carbon monoxide, and leaves the gasifier together with the products of the pyrolysis stage and steam from the drying stage.

Syngas can then be used to produce power, thermal energy or a range of chemical feedstock for industries, including liquid and gaseous transport fuels. Syngas can also be utilized in producing methanol, which is a useful fuel for heat engines. Another way of utilizing syngas is burning it directly in the open air, much like Liquid Petroleum Gas, and use for direct heat applications such as domestic or industrial applications (i.e., cooking, drying food and grain, boiling water, producing steam, greenhouse heating, and running of absorption refrigeration and cooling systems).

Syngas can be converted into liquid fuels through four methods; Fischer-Tropsch synthesis, methanol synthesis, mixed alcohols synthesis, and syngas fermentation. By Fischer-Tropsch synthesis coal-derived syngas and natural gas are converted into liquid fuels. Methanol synthesis is utilized to generate methanol by converting syngas, which is obtained from steam reformed natural gas or syngas from coal, whereas mixed alcohols synthesis generates a mixture of ethanol, methanol, butanol, propanol, and some heavier alcohols. Through syngas fermentation, which is a biological process different from the above-mentioned three chemical catalytic processes, syngas is fermented by anaerobic microorganisms and ethanol or other chemicals produced. Various process parameters affect the composition and calorific value of producer gas, such as biomass sources and gasifier design. A flow diagram regarding to a generic gasification plant is shown in Fig. 2.2.

Technology: Gasifiers can be classified into two broad categories; dense phase reactors and lean phase reactors, depending upon the ratio of the solid content of biomass fuel a gasifier can burn to the total volume available, which is called as density factor.

The feedstock fills most of the space in *dense phase reactors*. There are three types of dense phase reactors, which are commonly available in different designs, based on the operating conditions: downdraft, updraft, and cross-draft. In the *downdraft (also known as co-current) gasifier*, both air and fuel move in the same direction, downwards. It is the most common type of gasifier. The distinctly defined zones for different stages of the gasification process are aligned as; drying, pyrolysis, combustion, and reduction from top to bottom. Fuel is fed from the top and flows through the combustion and reduction zones. The main advantage of downdraft gasifier is, it is so designed that tar moves from the pyrolysis zone to the combustion zone where it is burnt or broken down resulting in a relatively clean gases mixture with a low tar content in the exit stream. This clean gas mixture is suitable to be used in gas engines. In contrast to the downdraft gasifier, in

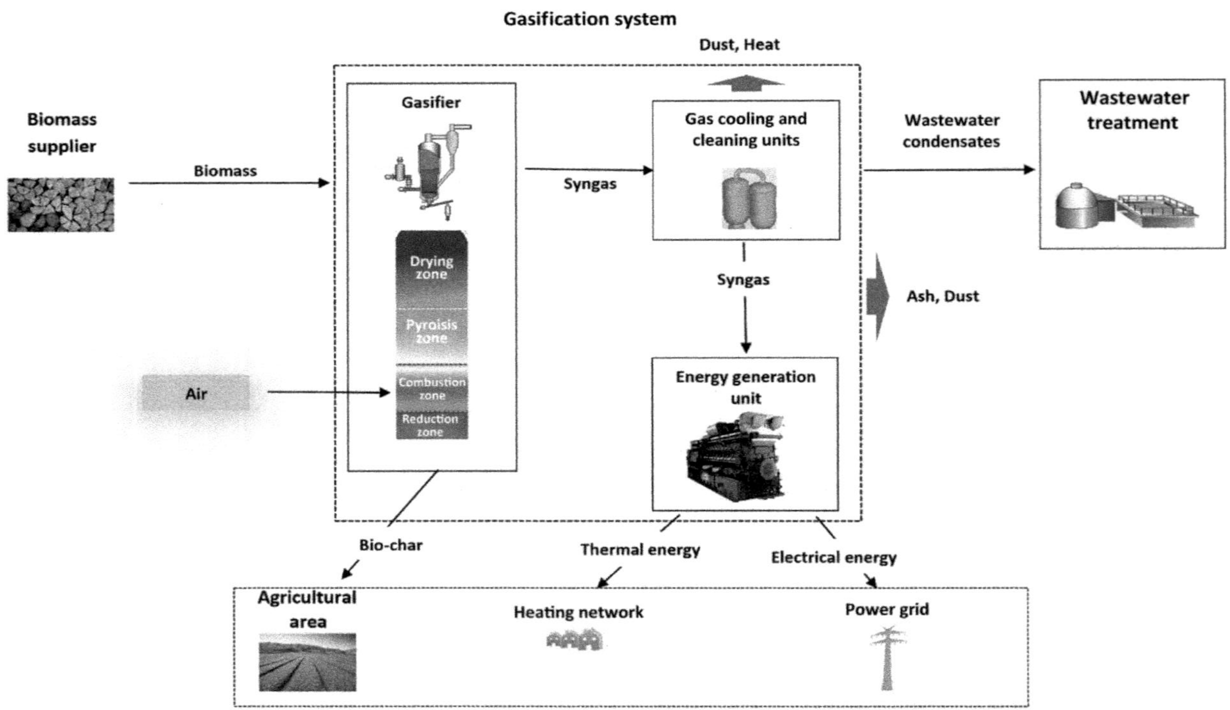

FIGURE 2.2 A flow diagram of a gasification plant.

the *updraft (or counter-current) gasifier*, air flows upwards from below the grate, against the fuel which moves downwards. In this type of gasifier, combustion takes place after the reduction phase where a producer gas, that has a higher calorific value because of its rich hydrocarbon (tar) content, is obtained. This producer gas is more suitable to be used for heating purposes, for example, in industrial furnaces, where it needs to be thoroughly cleaned to be used in power generation. In the third type of dense phase gasifiers, namely the cross-draft gasifier, air enters from one side of the gasifier and leaves from the other side. In *cross-draft gasifiers*, the residual ash does not get in the way of normal operation, but falls to the bottom of the reactor, hence they do not need a grate.

In *lean phase gasifiers*, all gasification reactions (drying, pyrolysis, combustion, and reduction) proceed in one large reactor chamber because they do not have separate zones for different reactions. There are two common types of lean phase reactors; fluidized bed gasifiers and entrained-flow gasifiers. In *fluidized bed gasifiers*, there is an inert bed of fluidized material such as char or sand in which biomass fuel is introduced. Air is distributed at the bottom of the fluidized bed through nozzles, which make it works as a hot bed of inert material, of which particles are agitated constantly by air. The choice of feeding biomass either above-bed or directly into the bed, depends on the density and size of the fuel and how bed velocities impact the fuel. After drying and pyrolyzing reactions occur, driving-off all gaseous content of the biomass fuel, the fluidized bed oxidizes the remaining char to increase the temperature of the process for the drying and devolatilizing to continue reactions. Fluidized bed gasifiers can produce more heat in a shorter time, due to the abrasion between bed material and biomass fuel, in comparison with dense phase reactors. In *entrained-flow gasifiers*, fuel is carried by the air in the reactor from the top of the reactor where both fuel and air are fed into. The main advantage of entrained-flow gasifiers is that the resulting gas contains very little tar. The main advantages and disadvantages of different types of gasifiers are summarized in Table 2.1.

2.1.1.3 Pyrolysis

Pyrolysis (also known as thermal crackling) is thermochemical decomposition of biomass, in the absence of oxidizing agent (air or oxygen), to convert biomass into liquid, solid, and gaseous products, namely bio-oil, charcoal, and fuel gas, respectively. The conversion process is realized by heating the biomass, in the absence of air, to high temperatures. Pyrolysis is the main chemical reaction that occurs naturally as the initial stage of combustion and gasification processes in the first few seconds. The major difference between pyrolysis and combustion is that combustion is done in the presence of oxygen, whereas pyrolysis occurs in oxygen-deplete conditions.

TABLE 2.1 The Main Advantages and Disadvantages of Different Types of Gasifiers

Gasifier Type	Advantages	Disadvantages
Downdraft 	• Adaptability to heavy gas loads • Low tar at higher temperatures, which makes it a good option for gas engines • Relatively clean gas with fewer particles	• High ash and dust content • Not a good option for biomass with very small particle size • Restricted scale-up opportunity • Tall design height, hence may not be suitable for small-scale facilities • Low flexibility in biomass specifications
Updraft 	• High thermal and biomass to gas conversion efficiency • Adaptability to biomass with different sizes • High charcoal burnt-out and hence low slag formation • Simple design and ease of use	• Sensitivity to moisture and moisture content of biomass • High tar formation • Poor adaptation to heavy gas loads • Resulting gas should be extensively cleaned if used for power generation • Longer start-up time
Cross-draft 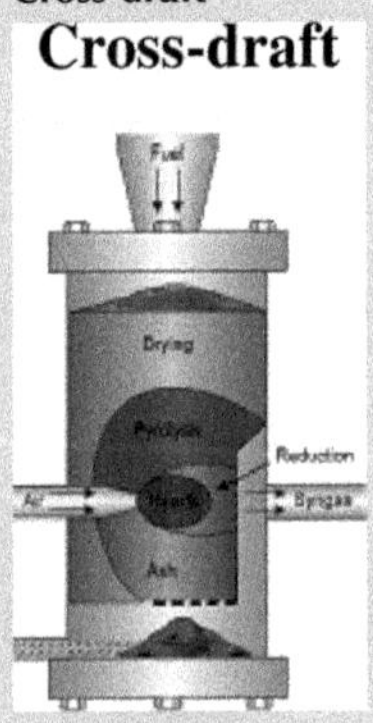	• Responsive to high loads and flexibility • Design tends to be small, hence suitable for small-scale facilities • Low cleaning requirements in resulting gas for high temperatures • Compact construction requirements	• High amount of tar and slag

(Continued)

TABLE 2.1 (Continued)

Gasifier Type	Advantages	Disadvantages
Fluidized bed 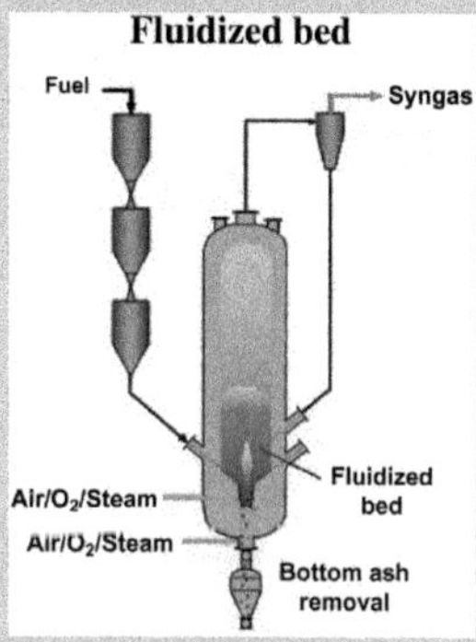	• Compact construction adaptable to small and large- scale facilities • Adaptability to biomass with different sizes • Low slag formation due to high ash melting point • Stable temperature pattern	• High amount of dust in the gas stream • Complexity of the system
Entrained-flow 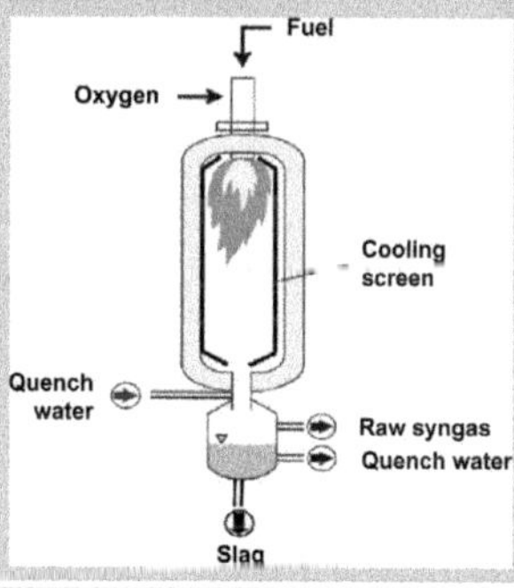	• Suitable for any type of fuel with a low moisture and ash content. • Adaptability to large-scale facilities • Requires short residence time	• High capital investment cost • Low flexibility in biomass specifications

Feedstock: A wide variety of biomass sources can be converted by pyrolysis-like energy crops such as switchgrass and miscanthus, forestry by-products, lumbering and timbering wastes, agricultural residues, and burned trees. Natural polymeric materials for the pyrolysis process can be assessed into three categories; cellulose, hemicellulose, and lignin. Smaller organic molecules (polymers) and inorganic compounds (minerals) are other components of the biomass suitable for pyrolysis. The distribution and proportions of these constituents in varying biomass types have a significant influence on the efficiency of the pyrolysis process as well as the composition and amount of pyrolysis products. For example, materials with high lignin content are especially appropriate for pyrolysis. Also, these contents contribute to different products of the process. For example, hemicellulose and cellulose components decompose to condensable vapors and gas, whereas primary products of lignin are liquid, gas, and solid char. Minerals remain in the char where they are named as ash and polymers contribute to liquid and gas products as well. Minerals may also have a catalytic effect on pyrolysis reactions that result in increased char yields.

Moisture content and particle size are the most important parameters related to biomass source on the process efficiency. Moisture content of the feedstock is an important factor that affects pyrolysis product yields. Ideally, it should be approximately 10% for process efficiency for fast pyrolysis processes (Bridgwater and Peacocke, 2000), whereas slow pyrolysis processes are more tolerant of moisture. A typical moisture content of 15%–20% is appropriate for wood in charcoal-making. Higher moisture content leads to high levels of water in the process, whereas lower levels may result in undesired dust production instead of oil. Feed particle size is another important parameter on the process yield that impacts the rates of char and liquid products. Biomass with larger particle sizes tend to give more char and most of the pyrolysis technologies are suitable for processing small particles because of the need for transferring the heat rapidly through the particle to maximize liquid yields. Hence, for a maximized yield of bio-oil, size of the feedstock has to be reduced prior to the pyrolysis process.

Process description: Production of bio-char and production of a dark brown liquid product, namely the bio-oil, constitutes the commercial applications of pyrolysis. The bio-oil obtained from the pyrolysis process is not suitable for direct use in internal combustion engines, however, it can be utilized as a feedstock in the gasification process for production of syngas and then bio-diesel, and as a substitute for fuel oil. Bio-oil is also an attractive option for co-firing because it can be handled and burned more readily than solid fuel such as waste wood and it is advantageous in storage and transport costs in that it has a much higher density than woody materials. Bio-oils have a wide range of utilization areas such as fertilizers, food flavorings, energy carriers for heat and power generation in engines, turbines, boilers, and furnaces. However, usage of bio-oil in fuel applications requires upgrading processes.

Bio-char, which is a useful by-product of the pyrolysis system, has a considerable potential in terms of carbon sequestration because of its ability to store the carbon dioxide in the soil resulting in a reduction of greenhouse gas (GHG) emissions to the atmosphere. This makes bio-char an efficient soil amendment that enhances the soil fertility to increase plant growth yield.

Pyrolysis typically takes place at temperatures between 400°C and 800°C. In this range, part of the lignin is decomposed along with most of the hemicellulose and cellulose contents in the biomass, and lighter molecules in the form of gas is obtained. Liquid bio-oil is obtained from the condensation of the vapors as these gases cool. The remaining part of the biomass, mainly composed of parts of the lignin, remains solid and forms bio-char. Hence, converting various kinds of biomass with different contents of lignin, hemicellulose, and cellulose, leads to different compositions of final products in terms of gases, condensable vapors, or the solid bio-char, to some extent.

The efficiency of the pyrolysis process in terms of product yields and their compositions are impacted by many factors such as reaction temperature, biomass composition, and particle size, heating rate, residence time, sweeping gas flow rate, and catalyst effect. Reaction temperature has a significant effect on product yields, in that increased temperature promotes further char decomposition, resulting in reduction in bio-char yields but enhanced gas yields. Size of biomass particles is an important factor on heat and mass transfer rates that directly affects the yield and characteristics of final bio-oil. Using biomass with larger particles leads to increased residence times of volatiles and hence higher yield of gaseous fraction, whereas smaller particle size results in enhanced bio-oil yield. Heating rate also influences heat transfer and different levels of heating rate have varying effects on product yield. For example, for obtaining maximum liquid yields along with a lower char yield, heating rate should be increased to some extent, to its optimum value. Sweeping gas flow rate also influences the residence time and hence the yield of the gaseous fraction of final products. Higher sweeping gas flow rate leads to increased yields of gaseous products, whereas char yields are decreased. Using suitable catalysts to regulate the process has also an influence on the constituents and quantity of pyrolysis products by affecting the decomposition behavior of biomass.

Pyrolysis processes can be classified into two main categories; slow pyrolysis and fast pyrolysis. Fast pyrolysis is currently the most commonly used pyrolysis system, which improved rapidly after the oil crises in the 1970s, and is designed so as to give a high yield of bio-oil from available renewable resources. Extremely high heating and heat transfer rates specify the characteristic features of fast pyrolysis as well as a rapid cooling of the pyrolysis vapors to obtain the bio-oil, a carefully controlled reaction temperature, and a very short residence time at the pyrolysis temperature (typically less than 1 second). At low temperatures (usually under 650°C), the yield of condensable vapors is maximized, as well as the liquid fraction of the products, whereas higher temperatures (above 700°C), maximize combustible gas yields. The specific features of fast pyrolysis generally require a biomass feed prepared as small particle sizes. The main product of fast pyrolysis is bio-oil and it takes seconds to complete the process, whereas slow pyrolysis takes several hours and results in bio-char as the major product. Pyrolysis temperature is a significant parameter that impacts the yield and composition of the bio-char. Long residence times maximize the bio-char yield, as well as low reaction temperature, in slow pyrolysis. The optimum temperature for slow pyrolysis is highly relevant to the purpose of the bio-char utilization. For example, to obtain bio-char with a higher volatile content, low process temperatures are appropriate. Besides these two common types of pyrolysis, there is also flash pyrolysis (i.e., very fast pyrolysis), which is typically used in the context of laboratory studies. The proportions of main products in flash pyrolysis are similar to those of fast pyrolysis, however flash pyrolysis applies higher temperatures and shorter residence times in comparison with fast pyrolysis, with the aim of preventing the condensation of the volatile products. In flash pyrolysis, the feedstock should be properly divided into smaller pieces for heating quickly in less than two seconds.

Torrefaction is another biomass conversion option, which can be seen as a mild form of pyrolysis at temperatures typically between 200°C and 320°C. Torrefaction is typically used for changing biomass properties by drying, prior to combustion and gasification applications, to provide a better fuel quality. Torrefaction can be combined with pelletization or briquetting to ease the transportation and storage of dense biomass in large-scale supply chains.

Technology: There are a number of technologies that are commonly used for the pyrolysis process including fixed beds, ablative systems, rotating cones, fluidized beds, circulating fluidized beds, and cyclonic systems, each representing a different reactor configuration. *Fixed beds* are generally used for the generation of charcoal with a slow heat transfer that leads to low-yield liquid for products. They comprise a cylindrical tube that contains a catalyst in pellet form in a static bed, through which biomass particles flow, for being converted into producer gas. The reactions occur while the gases and catalyst conduct each other. In an ideal reactor, a cooling system should be well-designed to prevent the deterioration and deactivation of the catalyst material in the bed due to excess heat that is not disposed from the reactor. Also, to control the process temperature in the reactor, fixed-bed reactors should contain internal tubes to transfer the heat. In the *ablative system*, there is a hot metal surface spinning, against which biomass particles are moved at high speed. In these systems, a high heat transfer rate is maintained by ablation of char formations at a particle's surface. Although this system prevents dilution of the products, it may present problems with its mechanical reliability. As an alternative to ablative systems, *cyclonic systems* include a heated wall, where the particles that are suspended in a carrier gas are introduced at high speed to dilute the pyrolysis products with the carrier gas. The major problem of ablative systems is that, their usage in large-scale commercial applications is limited because of the scale-up issues caused by a reactor design problem; while the reactor size is increased, the ratio of the wall surface to the reactor volume decreases, which causes technical complexities and decreases the efficiency of the pyrolysis process. In rotating cones, a mixture of preheated sand and biomass particles are fed into a rotating cone, in which the mixture flows. In *rotating cones*, inert heat carrier particles (generally preheated sand) and biomass particles are introduced to the bottom of a rotating cone. Due to the rotation of the cone, the mixture of heat carriers and biomass are converted by moving spirally upwards along the cone surface. *Fluidized beds*, which are beds of hot fluidized sand (generally fluidized by a recirculated product gas), are used as a medium for feeding biomass particles. In these systems, biomass particles are exposed to rapid heating due to high heat transfer rates from fluidized sand. Although some ablation occurs from the abrasion with sand particles, it is less effective than the ablative system. In *circulating fluidized beds*, biomass particles are fed into a circulating fluidized bed of hot sand, in which biomass particles move together with sand and recirculated product gas (or sometimes a combustion gas). As in regular fluidized beds, rapid heating of biomass particles is ensured by high heat transfer rates. In these systems, ablation also occurs and it is stronger than with fluidized beds. In contrast to ablative systems, regular and circulating fluidized bed systems can be easily scaled-up, however, they are rather complex and the final products are diluted, which makes it complicated to recover the liquid products. Hence, the quality and consistency of the final bio-oil should be improved after these processes.

2.1.1.4 Hydrothermal Liquefaction

HTL (also known as thermochemical liquefaction) is a conversion technology in which an organic crude oil is generated from biomass in a wet environment at a moderate temperature and high pressure. HTL typically aims at production of liquid bio-fuels for transportation applications.

Feedstock: A wide range of biomass feedstock can be converted into bio-oil via the HTL process. HTL is especially suited for the conversion of wet biomass feedstock in that the product yields decrease insignificantly when the biomass feedstock has more lignin content. Also, the feedstock with low moisture content, such as aged wood is less appropriate for the process. Some examples of biomass feedstock that are suitable for HTL process include, agricultural residues, forestry residues, wastes from food processing industry, grass, sewage sludge, aquatic energy crops, algae, and domestic waste with a high moisture content. The technical factors that may restrict the use of some biomass types in practical applications of HTL can be classified into two categories; criteria about biomass characteristics and criteria about reactor design. Criteria about feedstock properties include the quantity of contaminants such as rocks, sand, metals, water, and organic and mineral content in the biomass material. Size and homogeneity of biomass particles are other important factors in feedstock use. Type and ability of the pump and fluid dynamics are critical factors related to reactor design. Considering these criteria, some biomass types may require pretreatment before feeding into a reactor. The cellulose, hemicellulose, protein, and lignin content of the feedstock or feedstock mixture highly influence the quality and yield of the bio-oil. Many wet residue streams that contain a significant amount of contaminants such as garden, fruit, and vegetable residues may be currently available at low or even negative costs, however, they need expensive pretreatment to be prepared for HTL process.

Process description: Biomass feedstock with a high moisture content (preferably between 55% and 85%) is exposed to moderate temperature (in comparison with combustion and gasification processes) and high pressure in a liquid water environment, to obtain a crude-like oil (also known as bio-oil or bio-crude), by the HTL process. Before mixing with water, the biomass feedstock is preprocessed, to meet the size requirements and disposition of rocks, metals, and sand, to prepare the feedstock for the conversion process. The water–biomass mixture is then preheated and pressurized by a pump. In order to prepare a homogeneous mixture to prevent the problems with the flow of the pressurized feedstock in the pipes, special pretreatment and softening techniques can be applied at atmospheric pressure. The water–biomass mixture is then heated to the process temperature (250–550°C, generally 300–350°C) prior to feeding into the reactor, in which the thermochemical reactions takes place. These subsequent reactions decompose and convert the biomass mixture into a liquid effluent (a mixture of bio-oil and water) and some gaseous constituents by removing the oxygen content of the material in a significant amount as carbon dioxide and water. Before separating the bio-crude and water, the liquid effluent is cooled to prevent any reactions to degrade the bio-crude. The final bio-crude has a lower heating value and it's not miscible with water. Bio-crude can be separated in a light crude (LCR) and heavy crude (HCR) by extraction or distillation. LCR can be upgraded to produce transport diesel and HCR could directly be added to coal or replace coal to be co-fired in power plants.

The main factors that influence the efficiency of the process as the quantity of the final product are, characteristics of the feedstock, temperature and heating rate, pressure, residence time, and catalysts. Reaction temperature and heating rate play a critical role in specifying the production quantities of different phase streams. The optimal process temperature is highly dependent on the composition and characteristics of the biomass feedstock used. Char and gas formation increase when the process temperature is above the optimal value and/or the heating rates are too low, whereas temperatures lower than the optimal degree result in a reduction of overall product yields, and high heating rates lead to an increase in the production of bio-oil. Pressure is another important factor to determine the process yield, which plays a critical role in overall reaction kinetics. There is a significant relationship between residence time and feedstock characteristics as well as process temperature. If the duration of reaction exceeds the optimal residence time, further reactions occur, which is undesirable for an efficient conversion process. Water is the main catalyst in the HTL process. However, other catalysts such as water-soluble inorganic compounds and salts or transition metal catalysts can be included in the reaction to optimize the bio-oil yields. Due to the fact that the catalysts are able to convert various kinds of bio-materials with different contents into bio-oil, a wide range of biomass feedstock can be used in the processes that utilize catalysts.

Technology: After introduction of the concept of utilizing a mixture of hot water and alkali content to produce oil from biomass in the 1920s, HTL technologies attracted significant interest from researchers especially during the oil crisis in the 1970s. After a liquefaction process that takes place under high pressure was developed at the Pittsburgh Energy Research Center (Elliott, 2007), Shell Oil developed the Hydrothermal Upgrading process in The Netherlands (Toor et al., 2011). Today, technology companies in various countries such as Steeper Energy in Denmark and Canada, Mulchand Holdings in India, Licella/Ignite Energy Resources in Australia, and Altaca Energy in Turkey continue to explore the commercialization potential of HTL as an emerging oil production technology (Karatzos et al., 2014).

Four reactor types are commonly used for HTL applications throughout the world; plug flow reactors (PFRs), continuous stirred tank reactors (CSTR, fixed-bed reactors, and batch reactors. The liquid or slurry feedstock stream is continuously introduced into the *CSTR* (also known as completely mixed reactors) at steady-state conditions. The feedstock mixture is assumed to be uniformly distributed throughout the reactor and the product stream is removed simultaneously from the reactor. In an optimized reactor, the composition of the product stream removed from the reactor is the same as the liquid/slurry phase in the reactor and does not fluctuate with respect to time. CSTR technology is commonly used because it can be an established easily, in addition it is cheap and versatile. It provides a well-mixing of the feedstock stream in that the temperature and pH values are identical throughout the mixture and these values can be easily controlled during the reaction. In a *plug flow reactor* (PFR, in other words, continuous tubular reactor or piston flow reactor), biomass mixture is fed into the reactor continuously and moves through the reactor as a so called plug. PFR may either be one long reactor or comprise many short reactors in a tube bank in the form of continuous, flowing systems of cylindrical geometry. The concentration of the feedstock mixture changes while it travels along the long axis of the reactor. The dimensions of the reactor are key factors that determine the product yield. The *batch reactor* contains a tank with an agitator and integral jackets for heating and cooling the system. The biomass slurry is generally introduced via holes in the top of the reactor tank, through which vapors and gases also discharge. Bio-oil is usually taken out from the bottom of the tank. Another reactor type for HTL process is the *fixed-bed reactor*, which is explained in Section 1.1.2.

2.1.2 Biochemical Conversion Technologies

Biochemical conversion technologies conduct the breaking down of biomass into gaseous or liquid fuels through the use of microorganisms, bacteria, and enzymes. In this section, the most widely used biochemical conversion technologies are explained; anaerobic digestion (or biomethanation) and fermentation.

2.1.2.1 Anaerobic Digestion

Anaerobic digestion is a biochemical conversion process, by which microorganisms convert chemical energy in biomass feedstock into a gas with energy content in an oxygen-free environment. Anaerobic digestion has been applied commercially with success in a multitude of situations and for a variety of biomass sources. It is particularly suited for wet biomass (Faaij, 2006).

Feedstock: The feedstock for anaerobic digestion does not need to be waste, any feedstock with organic content (i.e., biodegradable content) can be converted into energy (or energy carriers) and digestate via the anaerobic digestion process. Non-woody plants, animal matter (generally manure and processing residues), sewage sludge, crop residues, municipal solid waste, energy crops, food scraps, and wastewater are among the typical feedstock for anaerobic digestion. Lignin content of biomass cannot be broken down by anaerobic microorganisms, hence using woody material, straw, and paper as feedstock in digester results in a slower digestion process. The composition, homogeneity, and biodegradability of the resources for anaerobic digestion vary considerably as there is a wide range of feedstock options. The following characteristics of biomass resources have a significant effect on the yield and quality of biogas product as well as bio-fertilizer; total solid content, degradation on the biomass during storage period, the purity of feedstock, the overall nutrient content and pH value. Typically, 70%–95% of the total solid content of biomass material constitutes biodegradable organic matter. For considering a feedstock as a valuable substrate for anaerobic digestion, its dry matter organic content should be more than 60%. The dry matter content of a biomass material is highly dependent on seasonal and operational conditions such as storage and pretreatment activities. The overall nutrient content of the substrate is another important consideration for bacterial biodegradation process. The degradation behavior of the feedstock is affected by the purity of the material, which as a result, affects the biogas quality and yield, hence the feedstock should be purified by removing unwanted inorganic matter such as metals, rocks, sand, glass, and plastics.

Process description: Biomass sources that are mixed in proper ratios are fed into a closed tank anaerobic digester with a loading rate in predetermined time periods. The conversion process is realized by anaerobic bacteria in multiple stages. Basically, there are three sets of bacteria that act in the conversion process. Firstly, bacteria break down carbohydrates in the biomass mixture to make them digestible by the second set of bacteria, which convert the resulting amino acids and sugars into hydrogen, carbon dioxide, organic acids, and ammonia. Eventually, the third set of bacteria generates methane and carbon dioxide from these products. Optimal temperature ranges should be defined for the growth of these mixed bacterial cultures.

After a time period called "retention time", a gaseous fuel, namely biogas occurs in the digester. The factors that affect the conversion efficiency in the digesters are given as follows:

- Length of the retention time: Length of the retention time depends on several factors such as biomass type, biomass amount to be converted, and process temperature.
- Process temperature: The digester can be operated at two different temperature ranges, namely mesophilic and thermophilic. Mesophilic digestion takes place optimally around 30–38°C, or at ambient temperatures between 20°C and 45°C. Thermophilic digestion takes place optimally around 49–57°C, or at elevated temperatures up to 70°C.
- Selection of the process temperature range is highly dependent on the available biomass sources, heating costs, and climatic conditions of the digestion site and digestate utilizaiton. Advantages of higher process temperatures arise from providing additional pathogen kill, however, they need more thermal energy for the digester to provide and maintain the appropriate temperature range. The lower temperatures can provide faster growth of bacteria that leads to a more robust bacteria population.
- Feeding rate: The digester feeding process can be batch or continuous. In a batch system, biomass is added to the reactor at the start of the process. The reactor is then sealed for the duration of the process. In continuous digestion processes, organic matter is constantly added or added in stages to the reactor.
- The total solid content of the biomass mixture in the digester: Dry digesters are designed to process materials with solids content between 25% and 40% whereas wet digesters can be designed to operate in a low solids concentration of less than 15%.
- The complexity of the system: The digestion system can have one or two stages. In a one-stage digestion system, all of the biological reactions occur within a single, sealed reactor or holding tank. In a two-stage digestion system, different digestion tanks are optimized to bring maximum control over the bacterial communities living within the digesters.

Biogas can be converted into electrical and thermal energy by the use of combined heat and power units (CHP) or can be utilized as a gaseous fuel without any conversion for domestic purposes such as heating or cooking. Digesters can be heated to their optimal operating temperature utilizing the thermal energy produced by CHP unit.

The composition of biogas does not meet the natural gas specifications in that the carbon dioxide and sulfur content in it is too high to be used as vehicle fuel. Biogas can be upgraded by removing its inherent carbon dioxide, sulfur, and moisture. This upgraded and less corrosive version of biogas is known as biomethane, which can be used as a valuable fuel to power motor vehicles for transportation. Biomethane can be further liquefied, and compressed, much like natural gas, and can be fed into the national natural gas grid. The liquefied biomethane can be more easily transported in comparison with the gaseous version.

Residue (or by-product) of the fermentation process (digestate), which is the solid remnant of the original input material to the digesters that the bacteria cannot use, can be potentially utilized in many places; as fertilizer in agricultural applications, as animal bedding, and as low-grade building products such as fiberboard. The most common application area of this by-product is the utilization as organic fertilizer in agricultural activities. The benefits of the utilization of digested biomass as fertilizer are twofold:

1. It is high-quality organic nutrient source that feeds plants and soil.
2. Long-term storage of the digested biomass increases the total cost of the system.

The digestate can be separated into its solid and liquid phase or can be used in croplands without separation as high-quality liquid fertilizer. If it is separated, the solid part (compost) can be used as a higher quality fertilizer with its higher nutritional value in greenhouse cultivation, horticulture, and high-quality pomiculture. A flow diagram regarding a generic anaerobic digestion plant is shown in Fig. 2.3.

Technology: The basic components/modules of a generic anaerobic digestion plant include (Karellas et al., 2010):

- a pretreatment module intended to collect, mix, and prepare the diverse feedstock and homogenize them in a single-feed stream;
- the digester itself, which is the key component of any anaerobic process operating;
- the gas treatment line, which constitutes biogas cleaning devices, the devices for the utilization of the energy content of the produced biogas;
- the solids-treatment line, to treat the digested biomass;
- a post-fermentation fertilizer tank to store the digestate;
- the biogas to energy generation unit, if the produced biogas is wanted to be converted into electrical and thermal energy.

A wide variety of anaerobic digesters are being utilized, which are the key components of anaerobic digestion processes, all performing the same basic functions in a different way, holding biomass in an oxygen-free environment and maintaining the appropriate conditions for the growth of the microorganisms that produce methane. Two main classifications can be considered to categorize anaerobic digesters: the first classification is based on the nature of the presence of the anaerobic bacteria in the digester (attached growth digesters and suspended growth digesters), whereas the second

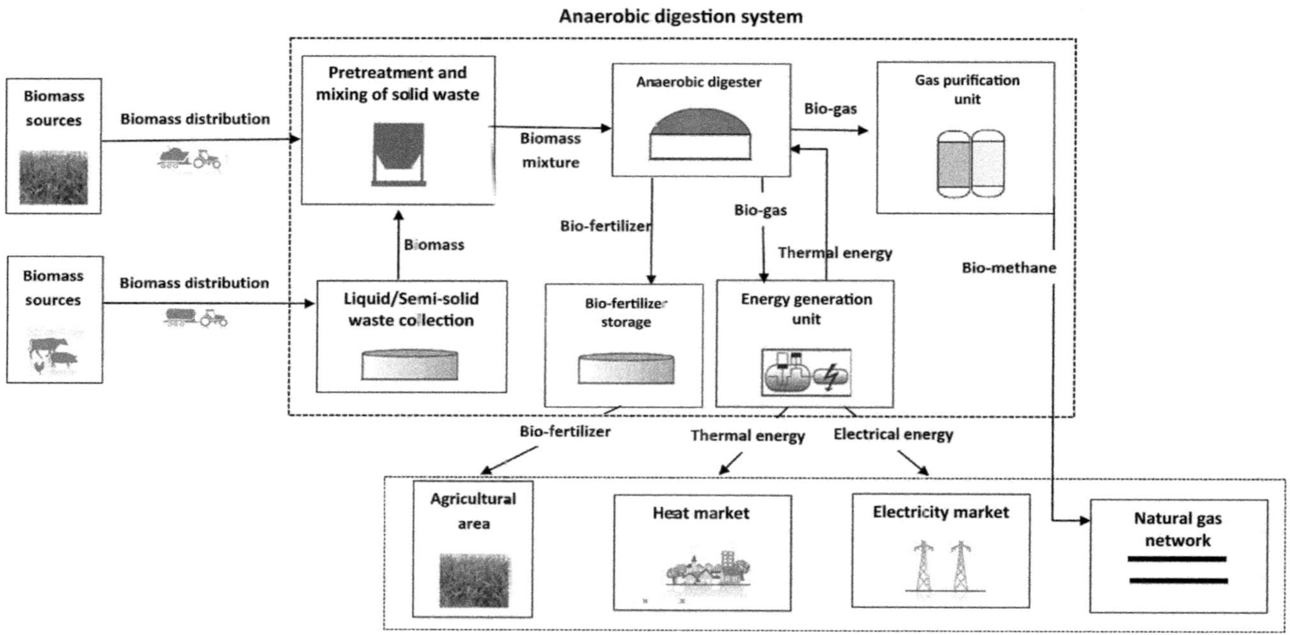

FIGURE 2.3 A flow diagram of a generic anaerobic digestion plant.

classification is done considering the organic loading rate (passive systems, low rate systems, and high rate systems). In attached growth digesters, methane-forming bacteria are fixed to a surface of a supporting medium where they grow and form biofilms. Microorganisms are suspended throughout the digester and mix freely with the biomass slurry in suspended growth digesters. The difference between the hydraulic retention time (HRT) and the sludge retention time (SRT) characterizes high-rate systems and low-rate systems. High-rate systems are specified by a relatively short HRT and long SRT, whereas low-rate systems have a longer HRT, which is equal to the SRT. High-rate systems are especially suitable for more heterogeneous slurries such as municipal wastewater, while low-rate systems can be used for treatment of all kinds of biodegradable resources (e.g., animal manure, organic industrial wastes, and organic fractions of mixed solid wastes). Examples of commonly used high-rate systems are, anaerobic baffled reactors, anaerobic filters (also known as fixed film digesters and attached growth digesters), sequencing batch digesters and upflow anaerobic sludge blanket digestion reactors (which are also suspended-growth digesters). Complete mix digesters, batch reactors, continuously stirred tank reactors, PFRs, and fed-batch reactors are among the typical low-rate anaerobic biogas digesters.

2.1.2.2 Fermentation

Fermentation is an anaerobic enzymatic conversion process including a series of chemical reactions to break down the glucose component of biomass material to its basic contents with the help of yeast or bacteria. Fermentation is used commercially on a large-scale in numerous countries to produce bioethanol from sugar crops and starch crops, which can be further distilled to obtain a purified alcohol to be used as an automotive fuel. The biomass is ground down and the starch converted by enzymes to sugars, with yeast then converting the sugars to ethanol (Mckendry, 2002). Fermentation involves the conversion of biomass by microorganisms in a process similar to anaerobic digestion but the products are alcohols or organic acids instead of methane.

Feedstock: Agricultural crops that contain high amounts of sugar such as corn, sweet potatoes, and sugarcane are efficient materials for fermentation as well as starchy materials such as rice, wheat, barley, and oats. Other biomass sources that are suitable for producing bioethanol include lignocellulosic biomass such as forest residues, agricultural wastes, food processing wastes, energy crops, wood residues, sawdust, and paper mill residues. Lignocellulosic biomass is basically made up of cellulose and hemicellulose contents that can be broken down into simple sugars, hence are fermentable, and lignin content that cannot be converted into sugars. Agricultural crops and lignocellulosic materials differentiate in the decomposition stage, however the fermentation, distillation, and dehydration stages are mainly identical for both of the biomass types. The yield from the fermentation process greatly depends on the proportions of cellulose, hemicellulose, and lignin parts of lignocellulosic materials, which changes according to the biomass source.

Process description: In the fermentation process, a variety of microorganisms such as bacteria, fungi, and yeasts, produce enzymes to catalyze chemical reactions for breaking down simple sugars into lower molecule sugars, which then be converted into chemicals and by-products. While these microorganisms consume simple sugars, to generate energy for their vital activities, they also produce by-products such as organic acids, carbon dioxide, hydrogen, and ethanol.

The fermentation process comprises multiple steps of chemical reactions, pretreatment and decomposition of the biomass, fermentation, recovery (by distillation, membrane separation, or pervaporation processes), and dehydration. Fermentable simple sugars are broken down and released by pretreatment and decomposition of the biomass material (i.e., cellulose and hemicellulose content in the lignocellulosic material). In the fermentation step, simple sugars are fermented by living organisms to produce organic acids or alcohols in temperature-controlled tanks. A fermentation broth is formed by mixing the sugars with water. The temperature, pH, and sugar concentration of the fermentation broth are important factors to meet the requirements of the microorganisms to facilitate the reactions. The action of the microorganisms is affected by possible impurities, as well as their own by-products. The fermentation process ends slowly, while the microorganisms digest all fermentable sugar or when fermentation by-products kill all of the microorganisms. As the product of the fermentation process has the ability of inhibiting the chemical reactions, the accumulated product should be removed during the process to increase the product formation. The recovery stage includes the purification and extraction of fermentation products of which concentration may be low in the fermentation broth that contains metabolic products such as soluble and insoluble medium components, living microorganisms, and cell fragments. In the recovery stage, the concentration of the fermentation product in the broth should be considered, as well as the intended use and desired properties of the product, to ensure good recovery or purification of the fermentation products. Centrifugation and filtration are the typical processes for the first step of the recovery to separate the large solid particles and microbial cells from the broth. Then, it is necessary to extract the broth into its major fractions. The major processes for fractionation include, reverse osmosis, ultrafiltration, gel filtration, supercritical fluid extraction, and two-phase aqueous extraction. Eventually, the final product is obtained by purifying the fraction that contains the product. To make the final product highly concentrated and completely clean from impurities, more precise chromatographic techniques and crystallization are applied.

However, each additional operation to further purify the product may result in some degradation, since none of these operations are fully efficient.

There is a wide range of enzymatic and acidic products through the commercial fermentation process in addition to bioethanol, including bulk enzymes, lactic acid, citric acid, and acetic acid. Also, intermediate chemicals that can be converted into various end products, such as succinic acid, butanol, and itaconic acid, can be produced by fermentation. Residues from remaining parts of cellulose and lignin can be used to generate electricity and steam in boilers. The excess wastewater from recover processes such as distillation and membrane separation should be disposed of properly as well as the residual solid materials, which can also be used as a soil amendment.

Technology: The main stages of the fermentation process (described above) may be carried out by separate equipment. Among major equipment, the pretreatment process may be realized using equipment for cleaning, purifying, and milling the feedstock. After pretreatment, the decomposition of simple sugars in a biomass resource may take place in a dilute-acid reactor to handle the hemicellulose content whereas a separate reactor may be used for enzymatic hydrolysis of the cellulose. Among commercially available hydrolysis techniques for the decomposition stage, dilute acid and concentrated acid are more traditional processes, whereas enzymatic hydrolysis has attracted recent interest of researchers. A relatively new research area is the co-fermentation of multiple sugar substrates, in the enzymatic hydrolysis of cellulose, by simultaneously performing saccharification and fermentation. By this process, hydrolysis of cellulose and fermenting microbes are combined, to handle the production of sugars and conversion of them into bioethanol by the fermentative organisms at the same time. Separate vessels may be used during the fermentation of five- and six-carbon sugars. There are mainly three technologies used for the recovery stage, distillation, membrane separation, and pervaporation. Distillation is used to separate two liquids with the help of the difference in their boiling temperatures, whereas a specific membrane is used for membrane separation to provide a barrier that limits the flow of some selected chemicals. Different methods are used to transfer the chemicals such as diffusion and convection. The differences between concentrations, pressure, or temperatures of chemicals are utilized in membrane separation. Pervaporation is a relatively new membrane-based technology to separate the volatile content from the liquid mixture. Using this system may lead to considerable amounts of energy savings in comparison with traditional distillation systems. The efficiency of the recovery process greatly depends on the duration of operation, which is characterized by the speed of the operation, due to the fragile nature of the product. Using the correct type and size of the processing equipment is also important to complete the process within time limits.

2.1.3 Biomass Extraction

Biomass extraction is a conversion process used to extract the oil of the seeds of various oleaginous plants. The residual solid of mechanical extraction process, namely "cake" can be used for livestock feed fodder. The reaction of bio-oil that is obtained from extraction process with alcohol leads to generation of bio-diesel. This process is termed as esterification.

Feedstock: The seeds of a wide range of oleaginous plants, such as cotton, soybean, groundnuts, sunflower, corn, and rapeseed, can be used as feedstock in extraction processes. To provide an efficient extraction process with high yields in oil, the oleiferous seeds should be pretreated by cleaning and screening before extraction or pressing. However, some seeds require further pretreatment besides cleaning. For example, grinding may be required to reduce their size. Heating, hammering, pulverizing, cracking, and extruding are among the other pretreatment options for some seeds prior to entering the press.

Rapeseed, soybean, sunflower, safflower, and corn are among the most commonly used feedstock for bio-diesel production. Besides agricultural crops, waste vegetable oils, and animal fat, especially tallow, can be used as raw material in bio-diesel production. In tropical regions, coconut trees, palm trees, and jatropha are utilized as well. The seeds should be stored in a cool and dry place and converted into oil as quickly as possible to prevent material degradation.

Process description: There are mainly two extraction processes commonly applied throughout the world: mechanical expression (also known as mechanical pressing) and solvent extraction processes. Depending on the type and content of the oil in biomass sources as well as the extraction degree needed to be applied, these processes may be applied consecutively or independently. By mechanical expression, the oleiferous seeds are exposed to mechanical pressure by permeable barriers so as to squeeze oil from the seeds or plant parts. The biomass material is placed and compressed in a container that is surrounded by small perforations from where the oil leaves the system. This process may either be batch or continuous. In a batch process, the oil is extracted batch by batch, in other words, one batch of seeds is pressed completely and then the container is cleaned before starting another batch, whereas, in a continuous process, the seeds are continuously fed into the container and exposed to pressure to pull the oil out of the seeds. Mechanical pressing may be followed by solvent extraction (also known as chemical extraction) in the cases where mechanical pressing is insufficient to squeeze all the oil from the biomass material. In addition, high-quality oil is obtained by mechanical expression; however, the yields are relatively low, so it is generally used in small-scale plants for producing special products or as a pretreatment process in a larger capacity solvent extraction plant. By solvent extraction, the seeds are contacted with a solvent that is used to dissolve and remove the remaining oil from the seeds.

The yield and efficiency of the extraction process are related to press configuration, seed type, pretreatment of the seed, and process temperature. For both of the process types, temperature is an important parameter that affects the process efficiency. The increased temperatures may either result in a decrease in the viscosity of the oil, which provides fluidity of the oil, or destroy the structure of cells of the raw material that may cause cooking. For extracting the maximum amount of oil, the seeds may need to be extruded before a final pressing. After the extracted oil leaves, a residual solid remains in the container. This residual solid is called cake when the oil is extracted by mechanical expression, whereas it is termed as meal when the seeds are exposed to solvent extraction. The oil content in the cake makes it a valuable animal fodder. The oil and protein content of the cake changes according to the type of seed used as feedstock in the mechanical expression process. For example, oilseed cake offers a good source of protein for cattle. In some cases, the cake should be heated to eliminate the toxic components before use as animal fodder. Cake from some kinds of seeds (e.g., mustard seed) is not suitable for animal fodder because they contain some elements that are harmful to animal health. Instead, they can be used as a high-quality soil amendment.

Bio-diesel, which is an environmentally advantageous renewable fuel, has a wide application area especially in the transportation sector with its structure being quite similar to traditional (fossil fuel-based) diesel. Bio-diesel can be produced by further processing the oil supplied by seed extraction as well as waste cooking oil or animal oil/fats. Basically, the bio-oil is reacted with alcohol to obtain bio-diesel. This further treatment of oil is called *transesterification*, which includes the reaction of the oil or fat with an alcohol (typically methanol or ethanol) to generate mono-alkyl esters (i.e., bio-diesel) and crude glycerol. Using methanol for bio-diesel production results in higher yields, however, ethanol may be more preferable in practical cases due to its low cost. In the transesterification reaction, either acids or bases can be used as catalyzers due to the acid content of the oil. The final product of transesterification should be purified to remove the by-products such as soap, residual alcohol, and water as well as glycerol.

Technology: There are two common types of press for mechanical expression, the hydraulic press and the expeller press (also known as an extruder or screw press). The main differences between two types of press are that the expeller press has a continuous operation mode and higher yield. As a processor of solvent extraction, the expeller presses can be used to extract part of the oil by reducing the size of the oleiferous seed and hence increasing its permeability to enhance the diffusion of the solvent. The oil seeds are fed into the one side of the expeller press, which is a screw-type machine that applies continuous pressure to compress the feedstock, and products leave the system from the other side. The expeller press can be used for scale of the operation. The operation of the hydraulic press is similar to that of the expeller press. The hydraulic press mainly consists of two parts, the fixed part and the mobile part. The screw-like mobile part compresses the material, whereas the fixed part consists of a plug where compressed feedstock is placed.

Besides producing bio-diesel with the help of catalysts, some relatively new catalyst-free technologies can be used for transesterification. Among them, supercritical and ultrasonic reactors have attracted the attention of researchers in recent years. By supercritical reactors, transesterification can be performed at the supercritical state, at which the phase boundary between liquid and vapor starts to disappear. Oil or fat dissolves in methanol to form a single phase, during the supercritical state of methanol when transesterification occurs. The reaction is then completed in a few minutes without using catalysts. In ultrasonic reactors, ultrasonic waves are utilized to create a cavitation by continuously producing bubbles in the reaction mixture. This cavitation provides the mixing and heating of the reaction mixture required for the transesterification process.

2.2 RECENT TRENDS IN BIOMASS-BASED PRODUCTION TECHNOLOGIES

Increased attention to biomass-based production, due to the drawbacks of using fossil sources-based production technologies, has led to recent research efforts to develop novel technologies and processes. Among them, bio-hydrogen, an example of an advanced (third-generation) bio-fuel, is seen by many as the fuel of the future. The main reason for this is its high energy density and high efficiency as an energy carrier. In addition, bio-hydrogen production results in negligible GHG emissions and environmental pollution. As a third-generation bio-fuel, bio-hydrogen has several advantages over the first- and second-generation bio-fuels. For example, producing first-generation bio-fuels, which are made from food crops, would lead to competition between food and energy purposes and subsequent increases in food prices, whereas third-generation bio-fuels would not. Furthermore, advanced bio-fuels are efficient energy carriers that could capture energy from sunlight approximately 10 times more efficiently when compared to second-generation bio-fuels, which are generated from non-food crops or wastes.

Bio-hydrogen can be produced through several conversion pathways. One of them is production through the action of living organisms called microbes, which are kept in special bioreactors (i.e., microbial fuel cells) filled with the organic matter as nutrients, sunlight, and carbon dioxide that they need to grow and reproduce. An advantage of bio-hydrogen production by this method is that, collecting it from the bioreactor is easier than collection of other types of bio-fuels such as bio-oils from the microbial cells that require purification, which is a complex and expensive process. Another way of bio-hydrogen production is steam reforming, which is a process for production of hydrogen, carbon dioxide, or other useful products from fuels with hydrocarbon content such as natural gas. Steam reforming involves the reaction of steam and the fuel at high temperature levels. Bio-hydrogen can also be generated by steam reforming of biomethane generated by the anaerobic digestion process. The third option for bio-hydrogen production is fermentation of renewable sources by various strains of bacteria either in light or in the absence of light. Relatively novel techniques are also being developed to increase bio-hydrogen yields, including using algae under anaerobic conditions and the development of a cell-free enzymatic process (Rollin, 2013).

Bio-hydrogen can be used as transport fuel, like hydrogen produced from fossil fuels, such as steam reforming of natural gas. The technical feasibility has been demonstrated by developing prototype hydrogen vehicles, however, utilization of hydrogen in the transportation sector is still an issue, due to challenges regarding to the capital costs and relatively short service life of hydrogen fuel cells and distribution of hydrogen as a transport fuel. Furthermore, in-vehicle storage of hydrogen is also an issue, and extensive research is being conducted on chemical storage options.

The latest information on the development of bio-hydrogen technology can be provided from the Fuel Cells and Hydrogen Joint Undertaking, which is a partnership supporting research and development activities in hydrogen energy technologies throughout Europe. In addition, BioHydrogen Platform provides an environment for information-sharing of recent developments for biological production of hydrogen from biomass and enables access to various projects and activities on this topic that are conducted by The International Association of Hydrogen Energy and The Netherlands Biohydrogen Network.

Plasma gasification technology is another emerging conversion technology. The viability of using plasma technology is explored for the gasification of a number of biomass types including municipal solid waste and industrial waste. Plasma torches are utilized to convert organic material into syngas with hydrogen and carbon monoxide content through an extreme thermal process. Argon or nitrogen can be used as inert gas depending on the size of the torch, which is ionized by passing through the plasma generated by the electric arc between the two electrodes in the plasma torch. The temperature of the process specifies the forming gas and the structure of the plasma. Biomedical waste and hazardous materials can also be treated by plasma technology, however too much inorganic material in feedstock may increase slag as waste of the system and decrease syngas production yield. The slag, produced as waste of the system, can be exposed to a cleaning process to remove the metals and the recovered slug can be sold as a construction material. Nonetheless, this technology still needs some improvement to mitigate the disadvantages resulting from the large initial investment and operational costs and, high energy consumption.

The Fischer-Tropsch synthesis is another technology used, to produce compounds based on hydrocarbon chains by a collection of chemical reactions. This chain of reactions converts carbon monoxide and hydrogen contents in coal, natural gas, or biomass sources into liquid hydrocarbon-based fuels such as synthetic fuel and lubrication oil in both gas to liquid conversion and coal liquefaction systems. Fischer-Tropsch process is a relatively old and well-known technology in comparison with bio-hydrogen production and plasma gasification; however it still gains considerable attention from researchers. Many refinements and adjustments have been made to the technology, including reactor design and catalyst development, after the introduction of the technology by two German scientists, namely Franz Fischer and Hans Tropsch, in the 1920s. The biomass-based conversion processes incorporating Fischer-Tropsch synthesis mainly consists of two modules, the gasification module and the Fischer-Tropsch synthesis module. The gasification module includes biomass handling, heat recovery, syngas cleaning, and recovery processes. The Fischer-Tropsch synthesis module receives the clean syngas and converts it into wax, tail gas, hydrocarbon condensate, and water. The primary product, namely wax, is then converted into smaller molecular weight hydrocarbon liquids. The design of the reactor, in which Fischer-Tropsch synthesis is conducted is important, due to the presence of highly exothermic reactions that make heat removal critical. Different types of commercially available reactors can be used for Fischer-Tropsch reactions, such as fixed-bed reactors, fluidized-bed reactors, and slurry-bed reactors. Between them, the slurry-bed reactor is widely preferred for low temperature reactions and provides higher conversion efficiency and better temperature control. However, for having high throughputs from high temperature Fischer-Tropsch synthesis, fluidized-bed reactors can be used.

2.3 DISCUSSION

Non-renewable resources are used frequently throughout the world in production and manufacturing processes. This is due to several reasons such as, lack of awareness and acceptability of renewable sources by producers and consumers, the technological complexities and immaturities as well as the cost of technology for renewable-based production, and an inconsistent supply of renewable resources due to natural factors. However, biomass-based production has been proposed as a part of the solution to the increasing worldwide resource depletion, climate change, and environmental pollution caused by the use of non-renewable- based systems. Decision-maker(s) should have a comprehensive knowledge about available biomass-based production systems, processes, and technologies that are or can be utilized throughout the chain, to effectively design the configuration of biomass-based production chains and manage operational activities. Considering this fact, this chapter aims to provide a comprehensive understanding of various biomass conversion technologies that are commonly used in biomass supply chains, focusing on the information to be used in decision-making procedures for design and management of biomass-based production chains.

Thermochemical conversion systems require extensive interaction between the thermal and mass fluxes, which makes it important to accurately analyze the system conditions and needs and to design efficient systems. Considering the significant amounts of smoke, tar, and ash particles emitted, design and optimization of environmentally acceptable thermochemical conversion systems is critical in terms of the minimization of these emissions and accommodation of their potential harmful effects. Choice of the most suitable technology should be made considering the specific advantages and disadvantages of different thermochemical conversion technologies, to minimize the effect and severity of the possible operational problems that can be encountered in biomass conversion.

Bio-chemical conversion systems are promising solutions for the treatment of organic material, preventing pollution and leading to efficient energy production. They are attractive processes for especially farmers because of the possibility of treating farm waste properly by mixing it with other organic materials in a nearby plant. In this case, they will be able to benefit from heat and/or electricity production as well as utilization of the digestate as organic fertilizer in their agricultural activities. Also, they will take the advantage of short supply site–plant-demand, site transport distances, as well as low required storage capacities on the overall process economics.

REFERENCES

Bridgwater, A.V., Peacocke, G.V.C., 2000. Fast pyrolysis processes for biomass. Renew. Sust. Energ. Rev. 4 (1), 1–73.

Elliott, D.C., 2007. Historical developments in hydroprocessing bio-oils. Energy Fuel 21 (3), 1792–1815.

Faaij, A.P.C., 2006. Bio-energy in Europe: changing technology choices. Energy Policy 34, 322–342.

Karatzos S., McMillan J.D., Saddler J.N., 2014, The potential and challenges of drop-in biofuels, IEA Bio-energy Task 39.

Karellas, S., Boukis, I., Kontopoulos, G., 2010. Development of an investment decision tool for biogas production from agricultural waste. Renew. Sust. Energ. Rev. 14, 1273–1282.

Mckendry, P., 2002. Energy production from biomass (part 2): conversion technologies. Bioresour. Technol. 83, 47–54.

Puig-Arnavat, M., Bruno, J.C., Coronas, A., 2010. Review and analysis of biomass gasification models. Renew. Sust. Energ. Rev. 14, 2841–2851.

Rollin, J.A., 2013. Synthetic enzymatic pathway conversion of cellulosic biomass to hydrogen, Doctor of Philosophy Thesis. Virginia Polytechnic Institute and State University, Blacksburg, VA.

Toor, S.S., Rosendahl, L., Rudolf, A., 2011. Hydrothermal liquefaction of biomass: a review of subcritical water technologies. Energy 36 (5), 2328–2342.

Chapter 3

Biomass-Based Supply Chains and Logistics Networks

Chapter Outline

3.1 General Structures and Key Components of Biomass-Based Supply Chains 55
3.1.1 Biomass Cultivation, Harvesting, Collection, and Storage 58
3.1.2 Biomass Pretreatment 59
3.1.3 Biomass Transportation 59
3.1.4 Biomass Conversion 67
3.1.5 Bio-Fuel Storage, Transportation, and Consumption 67
3.1.6 Energy Production and Transmission 72
3.2 Discussion 73
References 74

3.1 GENERAL STRUCTURES AND KEY COMPONENTS OF BIOMASS-BASED SUPPLY CHAINS

The supply chain is the network of organizations, people, and activities to transform natural resources, raw materials, and components into a finished product or service and deliver the product/service from supplier to consumer. Supply chains involve various stakeholders that include:

- component/raw material suppliers,
- storage facilities,
- distributors,
- manufacturing facilities,
- consumers.

Supply chain management involves coordinating and integrating the flows both within and between the stages of the supply chain. Three main flows in a supply chain can be described as follows:

1. The material flow, which involves the movement of materials, products, or services from supplier to end user. Customer returns or service needs are also included in the product flow.
2. The information flow, which provides the information transfer between supply chain components. Information flow involves transmitting orders and updating the status of delivery.

Decision-Making for Biomass-Based Production Chains. DOI: https://doi.org/10.1016/B978-0-12-814278-3.00003-0

3. The financial flow, which consists of credit terms, payment schedules, and consignment and title ownership arrangements.

To coordinate the flows and collaboration between the organizations in the most efficient ways, supply chain activities should cover everything from feedstock supply, product development, production, and logistics, as well as the information systems needed to coordinate these activities.

According to Papapostolou et al. (2011), in the context of the integrated biomass supply chain design, various problems emerge and call for solutions. More specifically, in strategic decision-making, the identification of the best supply chain configuration includes:

- the selection of highly productive raw materials and the decision on domestic cultivation or import;
- the coordination of transportation, pretreatment, and storage at operational, tactical, and strategic levels;
- the use of efficient biomass to energy conversion technologies to enable relevant reductions in environmental and biomass production costs;
- the location and capacity selection for the conversion facilities and storages;
- the partial or total satisfaction of the domestic demand from own production;
- the design and set-up of the storage and distribution network.

Management of biomass-based production chains comprises three interconnected and interdependent processes; production planning, inventory management, and transportation and distribution planning. Production planning involves the planning and control of cultivation, harvesting, collection, and pretreatment of biomass sources, conversion processes for production, and treatment of bio-products. Inventory management refers to all planning and management activities for storage of biomass sources and bio-products throughout the chain. Finally, transportation and distribution planning includes the decisions regarding to transportation and transshipment of; (1) biomass feedstock between supply sites and conversion, pretreatment, storage facilities; (2) bio-products between conversion, treatment, storage facilities, and consumption nodes. Fig. 3.1 depicts the process flow in a typical biomass-based energy production chain.

Biomass-based production chains incorporate three stages, namely the upstream stage, the midstream stage and the downstream stage, as illustrated in Fig. 3.1. The upstream stage includes the upstream of a production unit, from the extraction of biomass sources to the main conversion process (e.g., the conversion of biomass into fuel, energy, or other bio-products), whereas the downstream stage covers the part of the chain from the main conversion process to the product use and disposal phase. The activities related to the main conversion process form the midstream stage. The next sections explain the activities and processes covered in each stage.

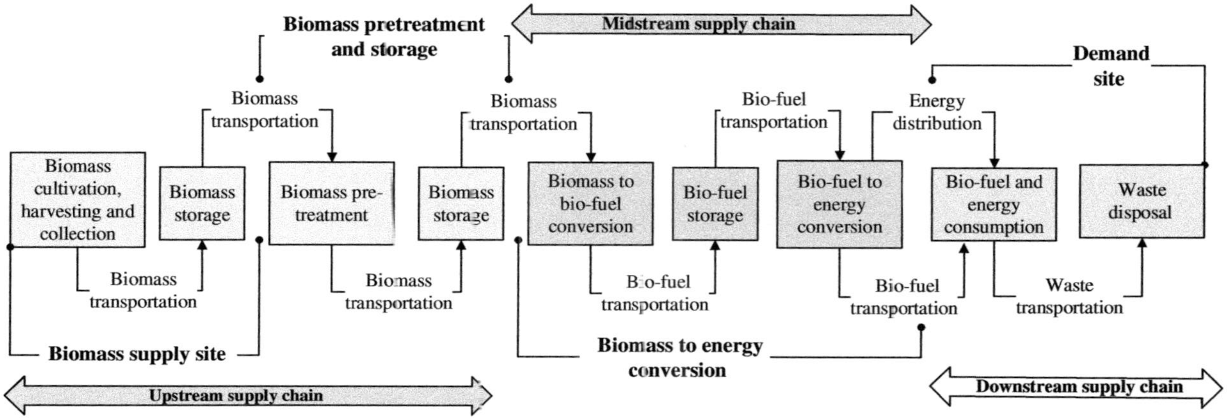

FIGURE 3.1 The process flow in a typical biomass-based energy production chain.

3.1.1 Biomass Cultivation, Harvesting, Collection, and Storage

Plant-based biomass sources need to be planted, cultivated, and harvested considering a growing cycle. Crop residues are usually collected after the harvest of the agricultural crops. The main decisions in biomass cultivation, harvesting, and collection activities are size optimization and allocation of agricultural lands, harvest scheduling, and biomass collection planning accounting for deterioration, seasonality, and availability of biomass materials as well as other biomass characteristics, climatic conditions, land availability, and bio-energy demand.

An optimized system for biomass cultivation, harvesting, and collection is essential for cost-effective and time-sensitive pretreatment, storage, and transportation of the biomass for conversion. The harvesting timing and frequency may affect the yields of energy crops, thus careful planning and scheduling would be necessary in order to guarantee the quantity and quality of the biomass supply (Sokhansanj et al., 2006). Although the biomass supply is usually discrete due to the seasonal availability, the bio-energy demand is year round. Therefore, the resulting operational challenge is to manage the biomass storage in order to maintain a stable supply for the production at conversion facilities. The major decisions related to biomass storage are storage mode selection, site selection, and capacity planning for the storage facilities. These decisions are influenced by biomass source types and characteristics as well as transportation options.

Biomass sources can be stored in various ways. Ambient and hypethral storages lead to a significant reduction in biomass to energy supply chain costs. However, aerifying (contingence with air) may cause deterioration of the biomass, as well as potential environmental and health risks. If a higher quality biomass supply is required, using covered storages with hot air drying capability can be employed to reduce the deterioration of the biomass. In this scenario, costs of biomass storage and handling will be higher but the material loss can be assumed to be negligible. As can be seen, the trade-off between storage cost and material loss has to be analyzed before determining the selection of storage modes.

Biomass storage can be centralized or decentralized. Centralized systems appear as limited storage accessible from multiple points within the supply chain network. As less storage is constructed in central locations in the supply chain network in these systems, investment and operational costs of storage will be reduced but high transportation costs may be endured. To reduce the transportation costs of the supply chain, the storage can be located in an intermodal point. An intermodal point can serve as an optimally planned location of centralized storage and distribution for biomass materials in which different transportation modes (road, train, etc.) meet.

Decentralized storage can be sited locally on-field, next to the conversion plants, which thereby reduce the costs of biomass delivery. Intermediate storage locations can be used for decentralized storage between biomass source sites and conversion plants. To facilitate the benefits of both centralized and decentralized storage, the decisions on distances and capacities should be focused on considering the trade-offs between storage and transportation costs. For detailed information on production, cultivation, harvesting, collection, and storage of different types of biomass resources, refer to Chapter 1 of this book.

3.1.2 Biomass Pretreatment

Raw biomass can be stored directly or pretreatment processes can be employed before the storage process. In the pretreatment process, after the biomass has been harvested and/or collected, thermal, mechanical and chemical treatments (or combination of them), are applied to increase its energy conversion rate and facilitate handling, storage, and transportation activities as well as to reduce the associated costs of the supply chain. With the pretreatment process, the size and moisture content of biomass is reduced to facilitate the transportation and contaminants are removed to improve the quality and stability of biomass feedstock.

Drying, carbonization, pelletization, chopping, shredding, grinding, and ammonia fiber expansion are among the widely used pretreatment techniques. The pretreatment method to be applied can be selected according to the characteristics of biomass feedstock. It should be mentioned that not all biomass materials need to undergo a pretreatment. For instance, maintaining a certain amount of moisture content in logs is considered a quality parameter (from a strength point of view), making them a good candidate for pelleting (Lehtikangas, 2001). For detailed information on treatment of different types of biomass resources, refer to Chapter 1 of this book.

3.1.3 Biomass Transportation

Transportation and associated logistics are major elements in a biomass-based supply chain that account for a large portion of the total biomass supply cost, energy consumption, and environmental emissions. Therefore, improvements in transportation and logistics operations will substantially improve the competitiveness of the biomass-based production sectors with nonrenewable-based production sectors. In biomass-based production chains, transportation planning mainly focuses on how and when to get biomass raw materials and distribute produced bio-fuel, from their respective origins to their destinations, and this has a significant influence on the overall costs of the supply chain. The decisions about major concerns of biomass and bio-fuel transportation such as distribution network configuration, distribution amounts, and transportation mode selection should be made optimally for cost- and time-effective supply chains.

Transportation is mainly concerned with loading and unloading operations as well as transferring biomass from one location to another. Transportation options for biomass to energy supply chains include roads, railways, waterways, pipeline, and/or a combination of two or multiple options. Selection of the transportation mode depends on the transportation distance, the form of biomass and bio-fuel to be distributed, and the transportation infrastructure available between the points of biomass source and destination. Environmental impacts should also be considered.

Road haulage (using trucks and tankers) are often used to ship biomass and liquid bio-fuel due to its ease of accessibility, operational flexibility, large transportation network, and high transportation capacity. Different from other transportation modes, highway infrastructure is larger and reaches many remote areas. Road transport is generally applied for relatively short distances when flexibility is required and multiple small farms have to be accessed, or when the infrastructure for other transportation modes does not exist. Road transportation is usually the cheapest mode of transport but it becomes more expensive than other methods as travel distance increases. Mahmudi and Flynn (2006) compared different transportation modes of delivering switchgrass (a kind of energy crop) biomass from farms to the bio-energy plants (e.g., maritime, road, rail, etc.) and concluded that road transportation using trucks is the most cost-effective transportation mode when the travel distance is less than 400 km. Total cost increases linearly with hauling distance, and decreases as truck capacity increases.

Rail transportation usually offers lower flexibility but has a lower per-ton-kilometer cost compared to road transportation for medium- to long overland transport distances. The advantages of rail transportation compared to maritime transportation are its operational flexibility and the rail network's larger size. Therefore, rail is the transportation mode of choice for long-distance and high-volume shipments of biomass.

Main maritime routes are composed of oceans, coasts, seas, lakes, rivers, and channels. Transportation by waterways is usually applied for international, high volume, and long-distance shipments. Waterway transportation requires high-capital investment but incurs low variable costs and lower carbon emissions compared to road and railway transportation options. In maritime transportation, barges have often been used to ship agricultural biomass. International liquid bio-fuel transports by waterways are usually performed with liquid tanks in vessels, based on the volume. It is required to establish special isolation or a heating system at extremely cold temperatures for transporting bio-fuels either in vessels, tankers, trucks, or rail.

Pipeline transportation is an alternative way to deliver biomass feedstock and bio-fuels. Pipeline routes are practically unlimited as they can be laid on land or under water. Although pipeline transportation requires high-capital investment, the large amounts of demand and supply to a large-scale bio-energy plant may justify the development of a pipeline network. Biomass slurry and liquid/gaseous bio-fuels can be distributed by pipelines with low per-unit operating costs. It has been shown that by using slurry from wood chips, pipelines would be economical in comparison to delivery by trucks only at large capacity and at medium- to long distances (Searcy et al., 2007). As the biomass feedstock mixes with the carrier fluid to ensure the fluidity in the pipeline, it is important to maintain the quality and stability of biomass. Also, it is required to provide a large amount of water resource. Pipeline use for transportation of biomass and bio-fuels may be restricted in some regions due to the necessity of maintaining pipeline temperatures in a specified range and prevent them from freezing. Biomass transportation by pipeline is the least-known technology and may prove to be the cheapest and safest mode of transport in the near future.

Intermodal transportation that combines two or more modes of transportation without changing the containment may be a good solution for large-scale biomass to energy conversion systems. Intermodal transportation may require the development of loading and unloading infrastructures of centralized storage or conversion facilities. Road transport and, for a few cases, train transport may be the only mode of transport. Combinations of waterborne and pipeline transport and train transport with road transport are commonly used for biomass and bio-fuels shipment. Trucks and tankers interface with trains at loading and unloading facilities. Waterborne and pipeline transport require interfacing with train and/or road transport at major facilities either on land or at the shore. Intermodal transportation facilitates the benefits of rail or waterborne transportation with low variable costs and road transportation with high operational flexibility.

In summary, the transport phase in the biomass to energy supply chain is associated with various transportation modes. Evaluation and analysis of alternative routes are therefore required to decide type, capacity, and schedule of the transportation, to minimize supply chain costs and travel time, as well as to minimize the environmental impacts of supply chain activities.

3.1.3.1 Transportation of Dedicated Energy Crops, Residues, and Wastes

In different supply chains, transportation costs are determined by the characteristics and operations at each step of the chain, such as the type of harvesting method (e.g., bale or chop), choice of preprocessing operation (e.g., compression, pelletization), storage method used (e.g., outdoor or indoor), and mode of transportation (e.g., truck, rail, ocean) (Yu et al., 2016). Average transport costs often include loading and unloading costs as well as transfer costs if multimodal transportation is the subject. The cost of chopping and compressing the material before distribution is also included in the case of wet storage. In addition, the biomass form, intended end use, supply, and demand locations (i.e., distance from the biomass source site to the storage or from the storage to conversion plant), and equipment and facility availability, have a significant effect on the performance and cost of the transportation system. Thus, the sustainability of the biomass distribution system requires optimized logistic chains, cost-effective transportation alternatives, standardized facilities, and infrastructure design between points of biomass dispatch and processing plant, optimized equipment configurations, efficient regulations, and environmental impact analysis (Miao et al., 2012). For a cost-competitive and reliable biomass transportation system not only delivery logistics, transport modes and route, but also the configuration of processing, handling, and transportation equipment and facilities should be optimized.

For biomass-based energy production, a large amount of biomass is required to be supplied to a commercial-scale conversion facility, given that biomass has a relatively lower mass and energy potential than other energy sources. Hence, locating biomass processing plants at the sites where the biomass resources are generated will be a good proposition for not bearing the additional cost of transporting biomass. However, in case of energy crop production, the areas that are uncultivated or are used for less transportation-intensive crop activities (i.e., pasture) are used for plantation of crops. Converting these lands into biomass production areas causes additional traffic on roadways that connect biomass source sites, storage, and the conversion facilities. In addition, as land price is a major component of energy crop production costs, commercial production of energy crops is most likely to occur in regions suitable to the high-yielding variety where also land costs are low, that is, outside industrial and residential areas. These lands may be at a certain distance, that is, in another region, from conversion facilities. Hence, in many cases for the supply of bioresources, long-distance transport may be necessary, which results in additional logistics costs, energy consumption, and ultimately higher greenhouse gas emissions compared to small-scale utilization. An industrial biomass-based production sector requires more truck traffic to supply and distribute biomass sources, which implies a potential environmental problem, increased GHG emissions from increased transportation activities between biomass source sites, storage, and conversion facilities. This environmental issue is presumably important because road transportation is a major source of GHG emissions (Fürst and Oberhofer, 2012). The International Energy Agency states that almost 25% of the energy-related carbon dioxide emissions worldwide result from transportation activities (IEA, 2009).

Large-scale conversion systems require greater use of biomass which results in larger scale feedstock handling and delivery infrastructure. In such systems, production/distribution centers can be utilized to facilitate biomass collection and transportation, where smaller quantities of biomass are consolidated, stored, and distributed to long-distance transportation systems. To reduce transportation costs by increasing truck payloads, especially for long-distance transportation, preprocessing equipment may be used to densify and reduce the size of biomass, which is normally collected in the form of bales from plantation sites, in these centers. In this case, trade-offs should be considered between preprocessing costs and transportation costs under various preprocessing scenarios. For example, lower transportation costs can be obtained when transporting a greater density/reduced volume of densified biomass, which requires higher preprocessing costs due to more complex preprocessing activities.

Globally, more commonly utilized options for transportation of lignocellulosic biomass resources are, road transport, rail transport, waterborne transport, and pipeline transport. Physical form, intended use, shape, quality, and quantity of the biomass, distance between supply and demand locations, capacity of the receiver plant, transportation facilities, and infrastructure have the greatest influence on the selection of the best transportation mode for the lowest delivered cost and environmental impact. Therefore, an evaluation and analysis of the economic and environmental performances of alternative modes, considering these attributes, is required by an appropriate framework to make decisions for transportation logistics and equipment configurations.

Road transport, which is generally well-developed and suitable for solid biomass distribution (i.e., energy crops), is usually the cheapest mode of transport but it becomes more expensive than other options as travel distance increases. Road transport is generally applied for relatively short distances (<100 km) when flexibility is required and multiple farm sites have to be accessed, or when rail and waterway infrastructure is absent. For example, in Austria, where typical road transportation of biomass for heating and combined heat and power (CHP) ranges from 20 to 120 km, tractor trailers are commonly used for short-distance transport (about 10 km) of unchopped thinning residues, forest wood chips, and various herbaceous feedstock (Hamelinck et al., 2005). Road transportation has low fixed costs compared to other transportation options, however, it has higher variable costs caused by fuel consumption, labor, tires, and wear costs. In addition, infrastructure limitations, traffic congestion, and environmental impact have to be considered, which result in indirect costs. For larger scale biomass systems, a larger collection area and longer transport distances are necessary to ensure year-round supply, which exacerbates these problems (Miao et al., 2012).

Trucks for transporting forest commodities and harvesting material are either tractor–trailer or fixed truck type. Tractor–trailers combine a road tractor and trailer and use standard highway road tractors, which are designed to pull cargo trailers. These trucks are designed for greater capacity and offer the versatility of changing the type, size, and configuration of cargo space. There are several trailer options available for hauling woody biomass. Among them, the most commonly used ones are log trailers, container trailers, and bulk vans. The primary aim of the log trailers is to transport trees, bundled materials, poles, or shortwood in racks. They have high payload capacities and mostly require unloading equipment at the end user's site. Bulk vans are primarily used for transporting forest and harvesting products. The total capacity of container trailers is less than log trailers and bulk vans. They mostly haul bulk material and can also be used as storage at the receiving facility. Container trailers are suitable for collecting materials in smaller volumes where road access is limited. Fixed trucks have a cargo area and are usually shorter than road tractor–trailer combinations, thus allow more maneuverability in pathways and tighter areas. Fixed trucks have lower payload capacity, which makes them work best when hauling distances are shorter.

Rail transportation is more cost-effective than road transportation when medium- to long-distance (>100 km) and large-scale biomass transportation is the case. Although it offers lower flexibility, and developing infrastructure for rail transportation requires a large initial investment cost, the lower variable cost, especially for transporting bales, logs, bundles, and industrial densified biomass, may make rail transportation a preferable option. Biomass type, quantity, form and distance from supply site to receiver facility are the factors that influence the cost of railcar equipment. The route infrastructure and transfer terminal policies specify the convenience of rail transportation for long-distance and large-scale biomass delivery. In most cases, multimodal road–rail transport, where trucks interface with trains at loading and unloading facilities of a depot or preprocessing facility, in particular, has been the focus for the bio-based industry (Flodén and Williamsson, 2016). The rail–road combination has also been found to have less environmental impact than competing modes (Kreutzberger et al., 2003). In addition, it is the most sustainable transport option when considering all three pillars of sustainability (economic, social, and environmental) (Flodén and Williamsson, 2016).

In long-distance and large-scale biomass transportation, waterborne transportation is another option. Similar to rail transportation, it requires high-capital investment for development and utilization of infrastructure (ports, ships, freighters, etc.), but incurs low variable costs and low energy use (Leboreiro and Hilaly, 2011). Waterborne transportation, which is especially relevant for delivery of pellets, wood chips, briquettes, and bales, is both used for international and inland deliveries, however in land-use, this transportation mode is restricted by the availability of waterways such as rivers and lakes. For instance, in Scandinavia it is relevant and common to utilize this mode, to transport the pellets for energy production by water, within Scandinavia as well as in Canada (Searcy et al., 2007; Mahmudi and Flynn, 2006).

Transportation of low energy density biomass feedstock via pipelines is the least-known and emergent option and may prove to be the cheapest and safest mode of transport in the near future. Pipeline transportation is generally associated with high-capital investment cost and detailed infrastructure works. But, using pipeline transportation for feedstock delivery to large-scale production units with large demand and supply rates may justify the establishment of a pipeline network with low per-unit operating costs. It has been shown that, in some circumstances, taking the advantage of the economy of scale of the plant, pipeline transportation is more economical than transporting by trucks, especially for medium- to long-distance distribution. Transporting a slurry of wood chips greater than 0.5 million dry tons per year for a one-way pipeline, and 1.25 million dry tons per year for a two-way pipeline, at medium- to long distances (greater than 75 km for one-way and 470 km for two-way), can be given as an example in such circumstances. Another example to consider is the transportation of corn stover (with 20% or higher solids concentration) by pipeline at capacities greater than 1.4 million dry tons per year (Searcy et al., 2007). As the transported material is mixed with the carrier fluid in the pipeline, maintaining the feedstock quality and stability may be a challenge in pipeline transportation, as well as maintaining pipeline temperatures and preventing them from freezing, especially in cold regions.

As transport distance and mode play a major role in energetic and environmental performance of the supply chain, the associated logistics network must be designed so as to ensure the best compromise between cost and environmental impact (Galvez et al., 2015). Large capacity production systems require developing a robust and integrated logistics system, where the bio-based resources and bio-products are shipped by different modes of transportation, so that an economic advantage can be provided over nonrenewable-sourced systems and the issues related to sustainability can be defeated. Adapting multimodal and co-modal transportation in supply chains offers opportunities to utilize the most appropriate mode, or combination of modes, and so increases the efficiency and decreases the environmental impact of a supply chain. Multimodality concerns the combination of different traffic modes in one transport chain and co-modality is the most appropriate use of each traffic mode (UN/ECE, 2001). The co-modality was a notion introduced by the European Commission in 2006 and refers to the "intelligent use of different modes on their own and in combination" to obtain "an optimal utilization of resources", so that the overall journey is the most sustainable that it can be. Both approaches have received much political interest in recent years. Co-modality has two particular differences from multimodality: (1) it is used by a group or consortium of shippers in the chain, and (2) transportation modes are used in a smarter way to maximize the benefits of all modes, in terms of overall sustainability (Verweij, 2011).

Intermodal transportation, which combines multiple transportation types, may be a solution for a large-scale biomass-based production plant, where supply and demand amounts are large, and long-distance (international or interregional) biomass shipment is required. In intermodal transportation, two or more transportation modes are combined without changing the containment to distribute biomass from supply site to the final destination. Here, the establishment or usage of a facility or infrastructure may be required, such as centralized storage or depot facilities to be used as distribution centers. For barge and pipeline transportation, major facilities, either on land or at the shores, are required to be established or utilized for interfacing with train and/or truck transport. For collecting and distributing biomass from farms to a local storage or processing facility, trucks can be used for short-distance hauling. After possible short-term storage or preprocessing, the biomass can then be loaded onto railcars or ships and transported from the local facility, which is alongside a rail track or waterway, to the production plant. The final leg of transportation for local distribution, from production plant to end user, can again be carried out by road. Thus, combining two or more transportation modes, intermodal transportation takes advantage of high flexibility of road transportation for local distribution and the low variable costs for rail or waterborne transportation for long-distance transportation (Perlack et al., 2005). Despite all the advantages of intermodal transportation, managing all of the modes for one journey may be complicated because of the increased loading, unloading, and material transferring activities, and because it requires simultaneous and effective communication between various stakeholders.

Transportation costs of low-density and high-moisture agricultural straw residues are a major constraint to their use as an energy source. As a rule of thumb, transportation distances beyond a 25−50 km radius (depending on local infrastructure) are uneconomical. For long distances, agricultural residues could be compressed as bales or briquettes in the field, rendering transport to the site of use a viable option.

Forest residues with low density and thermal value have high transport costs, hence it is economical to reduce the biomass density in the forest. In many areas, there is an existing collection infrastructure for harvesting wood and companies that harvest wood generally own equipment and transportation options for collecting forest residues, which are the primary advantages of using forest residues for power generation. Hence, these residues can be collected at the point of processing, which makes them suitable and relatively inexpensive sources of energy. In addition, the residue bundles can be handled and transported by the same forestry equipment, generally by conventional forestry trucks, which minimizes the need for new machinery.

All types of waste must be collected and distributed by leak-proof, covered containers or vehicles, which should be cleaned and disinfected after every use. In addition, special care must be taken when loading and unloading waste to ensure that there is no escape into the environment. In case of animal waste, liquid manure is occasionally disposed of through hoses and pipelines that may stretch for a certain distance depending on the proximity to the destination plant. Using a pipeline system is the safest way of transporting liquid and semi-solid wastes due to a significantly less possibility of damage to roads and the natural environment compared to transportation via tankers.

3.1.4 Biomass Conversion

As stated previously, biomass feedstock can be converted to various energy forms like electrical energy, thermal energy, and energy carriers (liquid and gaseous bio-fuels), by different kinds of conversion processes. For detailed information on biomass conversion systems refer to Chapter 2 of this book. Before executing biomass conversion, decision-makers need to deal with decisions related to the location of conversion facilities, conversion technology selection, as well as capital and operational planning of the conversion facilities.

3.1.5 Bio-Fuel Storage, Transportation, and Consumption

The biomass conversion industry faces many uncertainties, for example, yield and moisture content of biomass sources change according to weather conditions and bio-fuel demand and price depend on the market environment. To overcome the fluctuations caused by uncertain market conditions, produced bio-fuel should be stored for later requirements. Special storage considerations must be taken into account for each type of bio-fuel, that is, acceptable storage tank materials include aluminum, carbon steel, and stainless steel, depending on the bio-fuel type to be stored. Underground storage tanks are widely used to store bio-fuels as well as petroleum-based fuels and hazardous substances.

3.1.5.1 Bio-Fuel Storage

The typical factors that cause degradation of bio-diesel during storage without proper attention are, oxidation, exposure to air, temperature, chemical contamination, contact with water, and microbial degradation. Oxidation of bio-diesel results in formation of sediments in the fuel, which then clog the fuel filters. During storage, bio-diesel can degrade and be contaminated, due to the presence of water in the tanks and pipes of the storage and distribution system, and/or high moisture level in the air. Since microbes easily grow in an environment with water and nitrogen, microbial activity is another key factor that bio-diesel is susceptible to. To prevent degradation in bio-diesel resulting from these three factors, storage conditions should be optimized. For example, using the storage containers that are made from bronze, copper, zinc, lead, brass, and tin should not be used to store bio-diesel. Instead, tanks made from stainless steel, aluminum, fluorinated polyethylene or polypropylene, or fiberglass may be chosen. In addition, when the storage tank that contains bio-diesel is exposed to sunlight, excess heat and oxygen, bio-diesel will degrade more rapidly. Adding biocides to the fuel can be a way of stopping microbial activities as they inhibit the growth of microbes. Storage containers should be well dried and cleaned before usage to prevent water accumulation. Since bio-diesel can absorb moisture from the air, it is nearly impossible to keep water completely out of the container. Hence, the best way to prevent bio-diesel from water-based degradation is to use the fuel as quickly as possible, ideally in a few months. Furthermore, bio-diesel has a tendency to freeze at higher temperatures than petroleum-based fuels. Therefore, storage tanks must be designed to cope with this tendency. The freezing point of bio-diesel changes according to the type of oil or fat used as feedstock in its production. The freezing temperature of bio-diesel from saturated fats is higher than the freezing temperature of bio-diesel from unsaturated fats.

As in the case of bio-diesel, contact with water in the storage container is a major problem in bioethanol storage. As water is fully miscible in ethanol, the water–ethanol mixture has the potential to separate from gasoline with which it may be mixed, resulting in degradation in the quality of fuel. Hence, the tanks must be thoroughly dried before loading the blends of bioethanol and gasoline, to prevent separation of the phases. Also, if the container is not properly cleaned and dried, the ethanol that mixes with the water bottoms will remove any sludge and contaminate the fuel, because of the solvent action of ethanol. The storage temperature impacts the water solubility, that is, increased blend temperature leads to increased water solubility. Storage containers that are made from zinc, lead, aluminum, and tin should be avoided for storing bioethanol and its blends. Steel tanks are compatible and can be used with all ethanol blends. The tank that ethanol is introduced into should not have any corrosion, otherwise the problem will get worse and will result in contamination of the fuel.

Although biogas (or biomethane) is typically used at the same time as it is produced, it can be stored when production amount exceeds the demand, either for later on-site requirements or before/after transferring to off-site distribution nodes. The most common types of biogas storage systems are; low-pressure systems, medium-pressure systems, and high-pressure systems. Biogas storages can be internal or external. Internal biogas storage is integrated into the digesters. In most cases, they cover the digester as gas holders with flexible membrane materials. External biogas storage tanks are separated from the digester forming autonomous components of the anaerobic digestion plant.

Low-pressure internal systems are generally preferred for on-site utilization due to their simplicity in use and low cost. Floating gas holders, which can be used with a separate tank for the storage of the digestate, are widely used options for low-pressure storage of biogas, and which are made from fiberglass, steel, or flexible fabric. Flexible inflatable fabric tops are generally used with complete-mix digesters and plug-flow digesters. They are advantageous in terms of their low cost and safety, as they do not react with hydrogen sulfide in the biogas. The hydrogen sulfide and water content of biogas should be removed by cleaning processes for medium-pressure storage of biogas, to prevent the corrosion of the tank, and then the cleaned biogas should be compressed before feeding into the storage tank. Cleaning and scrubbing the biogas are also important for storage at high pressures, because the condensation and corrosion possibility due to impurities such as water and hydrogen sulfide is higher. Upgrading biogas so as to produce biomethane is a way to lessen the corrosiveness of biogas. Since biomethane is less corrosive than biogas, medium- and high-pressure systems are more suitable for storing biomethane. In addition, biomethane has a higher heat content than biogas and hence more valuable as a fuel, so that the investors may desire to endure the high-capital costs as well as high maintenance, safety, energy, and scrubbing needs of medium- and high-pressure systems for storing biomethane instead of biogas. Generally, biomethane is stored either as liquefied biomethane or compressed biomethane for future use. However, due to considerable evaporative losses, liquefied biomethane should be consumed quickly after production and stored for a week, at most.

3.1.5.2 Bio-Fuel Transportation

Transportation is among the highest of expenses to bio-fuel producers, and in most cases comes after feedstock and energy expenses. Storage requirements for bio-fuel influence the choice of appropriate transportation modes, that is, by road and rail transportation faster turnaround can be supported, while larger volumes of bio-fuel can be easily moved by barges. The market demand can be fulfilled more quickly by trucks, which is more adaptable to fluctuating market conditions, hence the storage needs may be lower. However, extensive cost savings may be achieved by transportation by barge due to large quantities transferred. Pipelines constitute a safe and reliable option for transportation of liquid bio-fuels. It is also economical for operating expenses and is environmentally favorable. However, the infrastructure cost is high but the current liquid pipeline infrastructure is suitable for transporting petroleum-based products in most parts of the world.

The ideal conditions for storage containers mentioned above are valid for the bio-diesel transportation container as well. The container should be made from stainless steel, aluminum, fluorinated polyethylene or polypropylene, or fiberglass and has to be cleaned and dried thoroughly prior to use. The container may need to be insulated due to significant changes in temperature, especially when transporting across regions or countries. Allowing the fuel to freeze in the container and then heating at the destination may be another way. The costs of tank insulation and heating the fuel should be compared with choosing the appropriate option. As pure bio-diesel is not flammable, warning signs may not be needed on the transportation vehicle (mostly tanker trucks), however, the blend of bio- and petro-diesel should be transported by a vehicle with a warning sign considering the flash point of the blend, which changes according to the bio-diesel content of the blend.

Since ethanol and ethanol–gasoline blends are classified as flammable liquids, tankers carrying ethanol and blends mostly display a placard with a flammable symbol. Although rail tanks are specified similarly, the different concentrations of ethanol and gasoline blends are labeled by different regulatory groups in most applications.

As biogas has a low heating value, it is less valuable as a fuel than biomethane. Hence, it is usually preferred to endure the transportation costs for distributing biomethane instead of biogas. However, in choosing the most appropriate form of fuel (biogas or biomethane), the cost of upgrading biogas into biomethane should be considered as well. The main advantage of liquefied biomethane is that it can be more easily transported. The transportation of liquefied biomethane is similar to that of liquefied natural gas, by insulated tanker trucks.

There are four main options to transport biomethane from its point of origin (usually production or storage facility) to the point of consumption (usually residential/industrial/agricultural area or fuel stations); distribution by dedicated biomethane pipelines, distribution by the natural gas pipeline, road transport of liquefied, or compressed biomethane.

When the distance to the destination from the origin is relatively close, especially for distances over privately owned property where easements are not required, the construction of infrastructure for dedicated above- or underground pipelines for biogas/biomethane transportation may be the most effective option. For longer distances, since biomethane has almost the same composition as natural gas, distribution through the natural gas pipeline network may be the option. In this situation, an unlimited distribution system is available, however, it is ensured that the composition and quality of biomethane should meet the specifications of the pipeline gas, and the biomethane producer should negotiate an agreement with the natural gas pipeline owner (private companies or municipalities) for injection into the grid.

In some cases, the distribution of biomethane through pipelines may not be practically or geographically possible or may be quite expensive. In these situations, road transportation of liquefied or compressed biomethane would be the solution. Compressing biomethane to high pressures is economically viable for road transportation as biomethane at ambient pressure has extremely low-energy density. Tube trailers can be used for road transportation of compressed biomethane (or similarly compressed natural gas). This type of transportation may not be a cost-effective option due to high-capital equipment and operational costs as well as the cost of further compression of the biomethane. Liquefying biomethane is an alternative way of facilitating biomethane transportation, however, there are some technical challenges due to the cryogenic form of liquefied biomethane. Liquefied biomethane can be transported by double-walled insulated steel tanks, which is also used to transport liquefied natural gas. The major advantage of road transportation is that the infrastructure is ready to be used.

3.1.5.3 Bio-Fuel Handling and Consumption

Although it can be used in pure form, blends of bio- and petro-diesel at any concentration are commonly consumed in diesel markets. Most countries use the coding system referred as the "B" factor to state the bio-diesel amount in any blend. For example, 100% bio-diesel is known as B100, whereas 20% bio-diesel, 80% petro-diesel is referred to as B20.

Blends that contain lower than or equal to 20% bio-diesel can be used in diesel engines with no or minimum modifications. Using pure diesel (100% bio-diesel) may cause some performance and maintenance problems in engines, hence some modifications may be needed. Blends can be made by mixing bio- and petro-diesel in tanks at the production node before loading to the transportation vehicle.

Due to its high biodegradability, and lower emissions, bio-diesel is the environment friendly alternative of petro-diesel. The major consumption area of bio-diesel is the transportation sector. It is widely utilized in road and rail transportation, however, in recent years applications to air transportation take place as well. In some countries, use of fuel from renewable sources in transportation is obliged by legislation, such as the Energy Policy Act of 2005 in the United States and the Renewable Transport Fuel Obligation in the United Kingdom. Besides use in the transportation sector, bio-diesel can also be used as, heating fuel in boilers for domestic and industrial applications, source in generators to provide energy in residential areas, and solvent to oil to extract oil spills from the shorelines (French McCay et al., 2004).

Bioethanol is an environment friendly renewable-based alternative to gasoline that can be used in petrol engines. As in the bio-diesel case, bioethanol can be blended with gasoline at any concentration. A coding system known as the "E" factor is used to state the bioethanol percentage in any ethanol fuel mixture in most countries throughout the world. For example, E85 depicts 85% ethanol and 15% gasoline. Blends that have lower ethanol content, typically from E5 to E25 are referred to as gasohol. Blends with a bioethanol content of maximum 15% are suitable for most of the existing gasoline engines. However, for the use of blends with high bioethanol concentration in the gasoline engines, minor modifications should be done. Bioethanol as a motor fuel in the transportation sector, mainly for road transportation, has been subject to many studies and much discussion (Larsen et al., 2009). Bioethanol can be used as fuel for power generation by cogeneration systems and thermal combustion. One of the major advantages of bioethanol is that it has higher octane content than neat gasoline, which enhances the compression of the engine and results in higher thermal efficiency. In addition, bioethanol is flueless, which makes it quite appropriate for using in residential areas. It is also used as raw material in the chemicals industry and as a solvent in the production of pharmaceuticals, detergents, cosmetics, coatings, and inks.

As previously mentioned, biomethane can be fed into the natural gas pipeline network since it substitutes natural gas, however, there are some barriers for injection of biomethane. The major barrier is the additional cost of biogas upgrading. The second challenge is about grid connection, that is, the cost of connection and the distance between the biomethane production site and natural gas grid. Some standards and certification schemes are proposed, by the governments of several EU countries, to define the restrictions about the composition of the biomethane to be injected into the natural gas grid. By these standards, the amount of the components, such as oxygen, sulfur, water, and other impurities, are established. Besides injection into the natural gas grid, biomethane can also be utilized in generating electricity and thermal energy in CHP units and as a power source in gas-powered vehicles at any ratio with natural gas. Biogas or biomethane can also be used for domestic purposes like cooking and heating, and in agricultural activities such as greenhouse heating.

3.1.6 Energy Production and Transmission

Bio-fuels can be converted into useful energy by using various equipments such as steam turbines, gas turbines, microturbines, fuel cells, reciprocating engines, turbo-generators, ignition engines, and cogeneration units. Cogeneration units, in other words, CHP units, simultaneously generate electrical and thermal energy. Typically, the heat generated in power plants, as a by-product of electrical energy production, is released to the atmosphere via cooling towers. However, this residual heat is recovered in CHP plants to be utilized in district heating systems. In modern cogeneration stations, approximately 85% of potential energy in biomass resources is converted into useful forms of energy, about 40% of the total energy production constitutes electrical energy, the remainder is being discharged as waste heat. Depending on the user's preference, it can be used to meet the internal heat requirement of the biomass-based production plant or for district heating with additional investment. The generated electrical energy is usually fed into the national electricity grid. Besides electrical and thermal energy, there is also the chance of simultaneous cooling generation by trigeneration or combined cooling, heat and power units.

Bio-energy plants with cogeneration units offer the possibility of installing district heating systems. District heating is a system for distributing heat generated in a centralized location for residential, agricultural, and commercial heating requirements such as space heating, greenhouse heating, and water heating. After generation, the heat is distributed from the energy station to customers via a network of insulated pipes. Usually, the pipes are installed underground, but there are also systems with overground pipes. The common medium used for heat distribution is water or pressurized hot water, but steam is also used.

The heat distribution system is made of mainly two parts that are required for the implementation of district heating of any size, that is, the main transfer station and the heat distribution system. The main transfer station includes the following major pieces of equipment: a heat exchanger where the heat required is transferred to the hot water that circulates in the distribution network, pumps that provide the energy to move the water in the network of pipes, and a data acquisition and control system. The cost of a heat distribution system depends on the length of the network and thermal energy flow in unit-time, where investment and operational costs of the main transfer station depend on the capacity of the system. Operational costs of a heat distribution system include, management, maintenance, and repair costs. Some of the heat produced by CHP units is needed for internal consumption during the fermentation process to satisfy the optimal process temperature for anaerobic digesters. Also, some heat is expected to be lost. To economically optimize the local heating system, it must be determined which of the heat customers should be connected to the grid. Within the system heat storage may be installed to even out peak-load demands. Thermal energy storage is a technology that stocks thermal energy by heating or cooling a storage medium so that the stored energy can be used at a later time for heating and cooling applications and power generation. The cost of thermal energy storage depends on its installed storage capacity.

3.2 DISCUSSION

The biomass-based supply chain is a special form of supply chain that includes biomass-based conversion facilities as manufacturing facilities. Biomass-based supply chains differ from traditional supply chains in several ways. For instance, supply of biomass sources is uncertain, seasonal, and constrained by land availability. Seasonality in biomass procurement induces uncertainties and fluctuations in energy production performance which forces transportation and storage activities to be managed more efficiently. In addition, supply chain design models for industrial products consider mainly demand (rather than supply), uncertainty, and seasonality, and focus on satisfying demand rather than making good use of the supply (Ekşioğlu et al., 2009). Development of a strong bio-energy sector depends on the development and management of biomass-based supply chains efficiently. Handling and delivery of biomass sources from the supply sites to the conversion facilities induce a variety of economic, energetic, and environmental implications. The main objectives of the biomass supply chain management are to minimize costs, environmental impacts of the supply chain, and ensure continuous feedstock supply (Gold and Seuring, 2011).

To perform the major activities in a biomass-based production chain, an efficient supply chain and logistics network has to be designed. Many decisions related to activities in various stages of the biomass-based supply chains involve trade-offs. For instance, locating the biomass-based production plants close to bio-product consumers will reduce the transportation cost of bio-fuel, but might increase the biomass transportation cost if the facilities are far away from the biomass supply regions. In addition, the trade-offs between transportation operating costs and fixed infrastructure costs can be critical for the profitability of the system. Due to the complex trade-offs involved, various competing supply chain network design decisions cannot be made independently. Therefore, comprehensive optimization and management of all individual components along the entire supply chain is essential to facilitate the economical, environmental, and social benefits of biomass-based systems.

REFERENCES

Ekşioğlu, S.D., Acharya, A., Leightley, L.E., Aror, S., 2009. Analyzing the design and management of biomass-to-biorefinery supply chain. Comput. Ind. Eng. 57, 1342–1352.

Flodén, J., Williamsson, J., 2016. Business models for sustainable bio-fuel transport: the potential for intermodal transport. J. Clean. Prod. 113, 426–437.

French McCay, D., Rowe, J.J., Whittier, N., Sankaranarayanan, S., Schmidt Etkin, D., 2004. Estimation of potential impacts and natural resource damages of oil. J. Hazard. Mater. 107, 11–25.

Fürst, E., Oberhofer, P., 2012. Greening road freight transport: evidence from an empirical project in Austria. J. Clean. Prod. 33, 67–73.

Galvez, D., Rakotondranaivo, A., Morel, L., et al., 2015. Reverse logistics network design for a biogas plant: an approach-based on MILP optimization and Analytical Hierarchical Process (AHP). J. Manuf. Syst. 37, 616–623.

Gold, S., Seuring, S., 2011. Supply chain and logistics issues of bio-energy production. J. Clean. Prod. 19, 32–42.

Hamelinck, C.N., Hooijdonk, G.V., Faaji, A.P.C., 2005. Ethanol from lignocellulosic biomass: techno-economic performance in short-, middle- and long-term. Biomass BioEnergy 28, 384–410.

International Energy Agency, 2009. Transport, energy and CO2: moving toward sustainability. International Energy Agency (IEA), Paris, France.

Kreutzberger, E., Macharis, C., Vereecken, L., Woxenius, J., 2003. Is intermodal freight transport more environmentally friendly than all-road freight transport? A review. In: 7th Nectar Conference "A New Millennium. Are Things the Same?", Umeå, Sweden, 13–15 June 2003.

Larsen, U., Johansen, T., Schramm, J., 2009. Ethanol as a fuel for road transportation main report, Technical University of Denmark.

Leboreiro, J., Hilaly, A.K., 2011. Biomass transportation model and optimum plant size for the production of ethanol. Bioresour. Technol. 102, 2712–2723.

Lehtikangas, P., 2001. Quality properties of pelletized sawdust, logging residues and bark. Biomass Bioenergy 20 (5), 351–360.

Mahmudi, H., Flynn, P.C., 2006. Rail vs truck transport of biomass. Appl. Biochem. Biotechnol. 129, 88–103.

Miao, Z., Shastri, Y., Grift, T.E., et al., 2012. Lignocellulosic biomass feedstock transportation alternatives, logistics, equipment configurations, and modeling. Biofuel, Bioprod Bior 6, 351–362.

Papapostolou, C., Kondili, E., Kaldellis, J.K., 2011. Development and implementation of an optimization model for bio-fuels supply chain. Energy 36, 6019–6026.

Perlack, R.D., Wright, L.L., Turhollow, A.F., et al., 2005. Biomass as feedstock for a bio-energy and bio-products industry: the technical feasibility of a billion-ton annual supply, Oak Ridge National Laboratory.

Searcy, E., Flynn, P., Ghafoori, E., Kumar, A., 2007. The relative cost of biomass energy transport. Appl. Biochem. Biotechnol. 136, 639–652.

Sokhansanj, S., Kumar, A., Turhollow, A.F., 2006. Development and implementation of integrated biomass supply analysis and logistics model (IBSAL). Biomass Bioenergy 30, 838–847.

UN/ECE, 2001. Working Party on the Transport of Dangerous Goods (WP.15). Available from: https://www.unece.org/trans/main/dgdb/wp15.html.

Verweij, K., 2011. Synchronic modalities–critical success factors. Logistics Yearbook Edition. pp. 75–88.

Yu, T.E., Larson, J.A., English, B.C., et al., 2016. Dedicated energy crop supply chair and associated feedstock transportation emissions: a case study of Tennessee. J. Transp. Res. Forum 55 (1), 51–66.

Chapter 4

Sustainability Issues in Biomass-Based Production Chains

Chapter Outline

4.1 Sustainability of Biomass-Based Production Chains 77
- 4.1.1 Economic Sustainability 78
- 4.1.2 Environmental Sustainability 82
- 4.1.3 Social Sustainability 86
- 4.1.4 Technical Sustainability 89

4.2 Design, Planning, and Implementation of Sustainable Biomass-Based Production Chains 91
- 4.2.1 Sustainability Indicators, Standards, and Certification Schemes 91
- 4.2.2 Sustainability Assessment in Design and Management of Biomass-Based Production Chains 100
- 4.2.3 Life Cycle Assessment of Biomass-Based Production Systems 105

4.3 Discussion 109

References 111

Further Reading 112

4.1 SUSTAINABILITY OF BIOMASS-BASED PRODUCTION CHAINS

The main idea of sustainability, sustainable production, and sustainable supply chain management is based on providing long-term benefits related to the use of scarce resources, through meeting today's environmental, social, economic, and other needs by efficient and responsible production and logistical activities that pay attention to economic growth, environmental protection, and standard of living simultaneously. Sustainability is the key point to prevent the depletion of resources that are needed for future generations. Environmental, economic, and social impacts of using nonrenewable resources and rising oil and energy prices have forced many countries to establish efficient strategies for further utilization of biomass resources, which are seen as one of the main diversification ways to reduce the dependence on fossil-based energy and thus, to improve reliability and security of

Decision-Making for Biomass-Based Production Chains. DOI: https://doi.org/10.1016/B978-0-12-814278-3.00004-2

energy supply. As an extension of such efforts, the 2002 World Summit on Sustainable Development adopted a Plan of Implementation to reach the goal of sustainable development. It strictly identifies the importance of "access to reliable, affordable, economically viable, socially acceptable, and environmentally sound energy services and resources" and in this context "promote a sustainable use of biomass".

Although as a rapidly developing sector, biomass-based production, trade, and utilization may well provide a promising supply option to meet increasing demand, many barriers may exist that can disturb or slow down a sound development of such systems, so that the utilization of renewable and biodegradable sources may not always be clean and sustainable in its production, conversion, and use. The barriers, which involve concerns about competition for land that may result in conflict with food production, natural resources, and biodiversity protection, as well as environmental impacts of increased biomass and bio-product transportation activities, especially in international trades, make sustainability an essential factor for long-term security and reliability of supply. Thus, unfavorable developments should be avoided by improving clear standards, measuring and assessing the harmonization of existing systems with these standards and developing assessment, design, and planning procedures to make choices of feedstocks, processes, and product mix efficiently, for well-designed biomass and bio-energy production systems, so that the problems in competition with other energy options in winning market share will be decreased by implementing biomass-based energy systems.

The key issues that promote sustainable operations in production systems and supply chains throughout the world are related to, fluctuations in feedstock and energy prices as a result of depletion of limited natural resources (especially, fossil-based resources), product liabilities, environmental pollution and global warming, waste disposal, quality of life, job creation, social responsibility and acceptability, economic growth and profitability, and development. It is necessary to consider and integrate the main pillars of sustainability, that is, economic, environmental, social, and technical sustainability, in design and management of operations and networks, to create sustainable production systems, supply chains, and logistics systems. Fig. 4.1 depicts the four pillars of sustainability and corresponding factors covered in this chapter.

4.1.1 Economic Sustainability

Designing, managing, and operating biomass-based production systems and supply chains in a way that it is economically sustainable means stimulating efficient and responsive use of all types of limited resources such as labor, capital, information, feedstock, natural resources, and energy. Economic sustainability is important in terms of providing long-term benefits from the resources, for continuity of the functions of business processes healthily over a number of years, and minimizing the economic risks that arise from

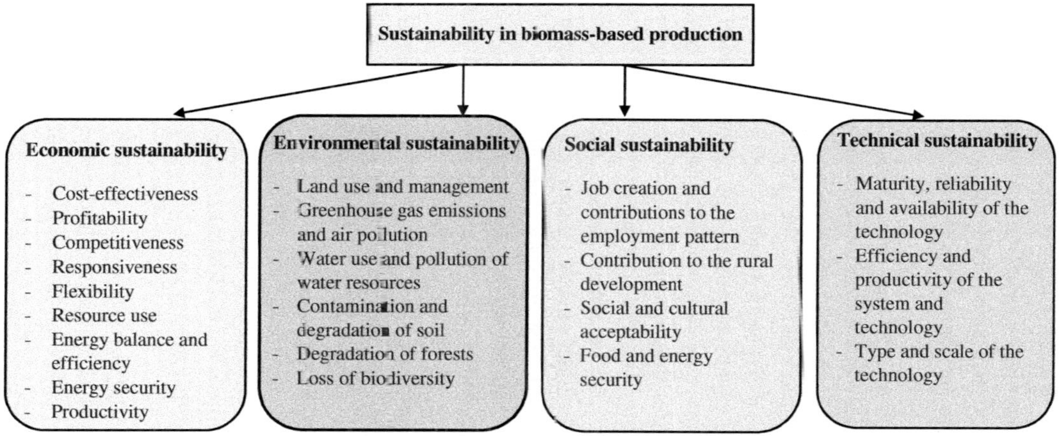

FIGURE 4.1 Four pillars of sustainability and corresponding factors covered in this chapter.

resource scarcity in especially profit-oriented businesses. As biomass-based fuel, chemical, and energy generation is becoming increasingly important, there is a foreseeable scenario of competition between the emerging biomass-based production sector and the traditional fossil source-based production throughout the world. To design, plan, and manage economically sustainable systems, it is essential to observe and analyze the economic performance of the biomass-based production chains. Economic performance of biomass-based production systems and supply chains can be measured by several metrics including cost-effectiveness, profitability, competitiveness, flexibility, and responsiveness.

Profitability, which can be used to evaluate the economic sustainability of both the entire supply chain and a particular unit of the chain, is a well-known and commonly used indicator of economic sustainability. Profitability is a measure of economic viability of the production chain, or is an element of the chain, and has a dynamic nature since product costs and prices are highly impacted by fluctuations in market conditions and policies. Especially for small- and medium-scale supply chains and plants, production and commercialization of useful by-products is important for increased profitability and economic viability. Besides, initiatives regulated by governments and certification schemes may be critical factors that contribute to profitability. Minimizing the unit production cost and/or maximizing the revenues from bio-product sales result in cost-effective operations and increased profitability, which effectively support the competitiveness and market acceptance of the products from biomass sources against fossil-based products, as well as the continuity of the supply chain functions.

Competitiveness, which can be seen as a critical factor for an enhanced economic sustainability, can be increased by participating in voluntary certification schemes, which may play a role in improving customers' perception about the biomass-based products that play a critical role in marketability of these products. Certification activities might also help in providing a price advantage to certified bio-products over their noncertified substitutes, which stimulates the competitive edge. However, additional costs related to applying and deserving certification should be considered in pricing a certified product.

Responsiveness is another important indicator of the economic performance of the biomass-based production chains that companies take into account, to gain competitive advantage over nonrenewable-based production systems. Responsiveness is the measure of meeting the demand of customers in short lead times with a high service level and in coping with a wide range of products. Responsiveness of the supply chain, which can also be defined as the measure of flexibility to the changes of market demand in a competitive and volatile business environment, can be considered as a factor to specify the design and management quality of the biomass-based production chain. To create a responsive biomass-based production chain, the common methods can be categorized into two; demand-oriented strategies and service level-oriented strategies. Maximizing the meeting rate of customers' demand and minimizing backlogs can be given as examples to demand-oriented strategies, whereas maximizing the service quality, minimizing the service time, and minimizing the lateness of product delivery to customers are the service level-oriented strategies. Responsiveness is important in terms of satisfying the expectations of delivery lead time sensitive customers. Besides responsiveness, resilience is an important factor for the economic well-being and sustainability of a supply chain. Hence, strategic decisions in design and management of a supply chain network and operations (including biomass supply) should be made, so as to adjust the tactical and operational decisions under possible disruptions, to maintain the critical functions and to ensure the continuity of the supply chain activities. In a resilient system, the uncertainties and risks associated with possible disruptions should be managed and mitigation strategies, including preventive actions, should be developed in advance to minimize the negative effects of a possible disruption, and contingency strategies should be taken when a disruption happens, with the aim of returning the supply chain to its original condition (Tomlin, 2006).

Resource use and energy efficiency and balance are also important indicators of economic performance of the production chain. The critical factors that impact energy efficiency of the system are the total energy demand and consumption throughout the supply chain as well as the proportion of fossil-based energy in total energy demand. Conventional means, vehicles and materials should be modified or replaced with their energy efficient substitutes in all elements of the biomass-based production chain, to become more energy efficient, so that less energy can be expended to produce the same amount of bio-products. Energy efficiency is highly related to the utilization of renewable sources for energy production and the integration of them would provide a sustainable energy policy throughout the supply chain.

Productivity is a measure of output from a production process, per unit of input, and can be used to measure the efficiency of the process by which inputs are converted into end products. Productivity of the system is primarily related to the resource availability and use efficiencies in biomass-based production, processing, storage, and distribution. To increase productivity of the system, the resources that are critical for the continuity of the production system (especially, nonrenewable resources) or supply chain, should be specified and their usage should be minimized, to optimize the input–output ratio, which is important to achieve the maximum utilization of available resources (e.g., raw material, labor). In a biomass-based production chain, productivity of biomass resources, the efficiency of biomass preprocessing, the efficiency of conversion of biomass resources to final bio-products, and the associated production costs per unit of production are critical factors for the overall productivity of the system.

A wisely designed biomass-based production chain for fuel and energy generation from local biomass sources has a considerable potential to increase the energy security of communities by providing sustainable energy supply from renewable sources. Therefore, the energy demand of the population can be fulfilled reliably by quickly responding to energy shocks, and long-term dependence on fuel and energy imports can be mitigated. Mainly four factors can be used to measure the potential of providing energy security of a biomass to fuel and energy production chain; (1) the consistency and reliability of procurement of feedstock, activities, and other resources that support the sustainable production of energy and fuel, (2) the consistency and reliability of product and by-product supply, (3) the stability of demand, and (4) flexibility and responsiveness of the feedstock supply and product distribution system. The consistency and reliability of supply of feedstock and other resources greatly depend on the price and biomass production volatility. Price volatility can be estimated by historical data, and fluctuations in biomass yield can be accommodated by increasing the availability of a diverse set of substantial biomass sources, flexible operation practices, and schedules into the system, due to unexpected stressors (e.g., floods, disease). Efficient management and operation practices for feedstock inventories are also important for a resilient feedstock supply. The consistency and reliability of energy and fuel products supply greatly depend on the management of product and feedstock inventories and reliable production/conversion and logistics processes. The stability of the fuel and energy market, governmental policies and initiatives have a significant impact on the demand fluctuations.

To override the hurdles that hinder the economic sustainability and widespread adoption of biomass-based production technologies, government policy support is considered as a key factor to moving commercial biomass-based production projects development forward. Financial incentives, created by governments to make clean and sustainable technology investments more attractive, support economic sustainability of commercial biomass-based production markets by increasing their profits, while competing with nonrenewable source-based markets where polluting technologies are used.

4.1.2 Environmental Sustainability

In recent decades, as a result of increased environmental awareness and stringent environmental regulations, environmental issues in supply chain design and management have received considerable attention from researchers and decision-makers in the biomass-based production sector. This consideration promotes the integration of various environmental factors in decision-making in biomass-based production chain design and management instead of focusing only on economic factors. Environmental sustainability is critical in the design and management of a supply chain network, by simultaneously handling land and natural resource utilization, environmental management, waste management, and pollution prevention practices. These practices are unequivocally essential to keep the environment clean, in which production, transportation, storage, and other supply chain activities take place, so that future generations can also meet their requirements and continue these activities in that environment. Wisely designed systems, supply chains, and logistics networks for biomass-based production can promote the intensified usage of biomass sources to produce fuel, energy, and other bio-products emphasizing the potential to reduce the environmental impacts of utilization of fossil-based sources. The main issues that practices for environmental sustainability of biomass-based production chains deal with are; land-use and land management, greenhouse gas (GHG) emissions and air pollution, water-use and pollution of water resources contamination and degradation of soil, degradation of forests, and loss of biodiversity.

Land-use and land management for biomass production purposes are other issues that stimulate food versus fuel debate, which is related to using the arable land for producing crops like soybeans, corn, and sugarcane for energy feedstock instead of food purposes. Such competition with food may result in decreased availability of foodstuffs and hence, increased food prices. For this reason, cultivation and consuming plants that cannot be used for food purposes, such as miscanthus, switchgrass, and algae are currently found to be more sustainable as they do not compete with food. Applications that require change in land-use, to produce resources for biomass-based production and land management issues, should be carefully handled in the design and management of biomass-based production changes using appropriate modeling and design methods. Land-use can be characterized by the land footprint, which can be defined as the total amount of land that is used to during production or development of a final product or service.

Environmental sustainability issues in the biomass-based production sector have gained considerable attention, since it is realized that more use of biomass-based sources in the production sector leads to more transportation and logistics activities, which result in more fossil fuel-based energy consumption and related GHG emissions. GHG emissions are commonly taken into account as an important environmental performance indicator in the studies and applications that focus on the environmental sustainability of biomass-based production chains. GHGs are atmospheric gas compounds that trap heat through emitting radiation in the atmosphere. The increase in the amount of these gases results in keeping the surface of the earth warmer by absorption of the sunlight that passes through the atmosphere freely. This situation then causes the greenhouse effect and enhances climate change as a result of global warming. In biomass-based production systems, the amount of GHG emissions to the atmosphere may be significantly lower compared to nonrenewable source-based production systems depending on the production and supply chain practices. For example, the use of fossil-based resources (e.g., coal, oil, gas) in fuel and energy production accelerates carbon (a highly effective GHG) emissions to the atmosphere, whereas the use of plant-based biomass resources in energy and fuel production has the potential to produce fewer GHG emissions since they are fully degradable, low-carbon and cleaner energy sources, due to the fact that carbon released into the air during energy generation process is sequestered during plant growth. However, there are two factors that disrupt carbon balance of plant-based biomass systems; firstly, the carbon balance can be achieved when the new plant generation has reached a harvestable size, which generally takes a long time; secondly, carbon emissions, which are due to the consumption of fossil fuel-based energy and fuel for farming practices that include production, harvesting, collection, handling, preprocessing, and transportation of biomass sources as well as production and distribution of bio-products, should be taken into account for a factual measurement of carbon emissions related to biomass-based production chains. To reduce these non-biogenic carbon emissions, farming methods can be improved by increasing the use of energy and fuel from renewable sources. Besides carbon dioxide, the emission of other harmful atmospheric gases and pollutants, such as non-carbon GHGs (e.g., water vapor, ozone nitrous oxide, and methane), should be carefully measured to monitor and analyze the environmental performance of the biomass-based production systems and supply chains. The performance of the supply chain in terms of total GHG emissions can be specified by the total carbon footprint of the system. Carbon footprint can be defined as a measure of the total amount of carbon dioxide and methane emissions caused by a system, event, or activity. The carbon footprint of a system is specified as carbon dioxide equivalent, however, it is generally a complex procedure to accurately calculate the total carbon footprint, due to the insufficient and unclear information about supply chain activities and elements, especially those that integrate processes using natural resources that capture or release carbon dioxide and/or methane.

The level of water-use throughout the supply chain and contamination of water resources that are close to the elements of the chain (i.e., production, storage, preprocessing, and distribution units) are also critical factors of environmental performance of biomass-based production systems. To keep the water resources fresh and clean, while performing supply chain activities, are essential for environmental sustainability and public health. Water-use can be defined as the amount of water consumed for a particular purpose in any point of the biomass-based production chains. Water-use in a system can be measured by its total water footprint, which can be defined as the total volume of water consumed, evaporated and/or polluted during production, consumption, and logistical activities, in a business environment. The supply chain design and management practices should be well-organized to allow the water resources to be consumed responsively and sustainably, while performing the major supply chain activities, in order to establish the current and future rates of use. This is especially important in cases where a significant quantity of water has to be used to have an efficient production process and high yields (e.g., anaerobic digestion process). Also, the impact of supply chain activities on the contamination of water resources, and accordingly on the economic and environmental well-being of the related area, should be efficiently assessed. By effective water management practices water usage can be balanced to meet both the current and future water demand of the production chain required for the sustainability of the production and other processes.

Soil contamination and degradation are two important considerations that affect soil formation and quality. Soil contamination can be defined as, the critical increase of chemicals and other toxic materials in the natural soil environment, while performing industrial, domestic, or agricultural activities, or by improper disposal of wastes. Soil degradation is evolution of the soil, due to local climatic conditions (e.g., rain, wind, and other man-made and natural phenomenon) and plantation practices, which results in the washing away and replacement of the equilibrium of a stable soil, and accordingly the modification of the composition, amount, and quality of the soil. Human activity is the major reason for soil contamination and degradation, which may also be perceived as undesirable ecological disturbance to the soil. During plantation, cultivation, and harvesting planning, necessary actions should be included into the planned operations for habitat protection and erosion control.

Another important environmental consideration in design, planning, and management of biomass-based production chains is maintaining the health of the forest ecosystem, which is especially critical for the cases where forest biomass (e.g., wood and residues from forestry operations) is used as feedstock in the conversion processes. There are mainly three functions of forest biomass in the forests; (1) naturally, forest biomass is an efficient fertilizer that helps to maintain the productivity of forest soil with its nutrient content, (2) it helps the forest species to survive by providing the nutrients and shelter they need, (3) forest biomass has a considerable impact on the quality, transfer, and distribution pathways of the water bodies in the forest, by affecting the ability of soil to hold and regulate water flows. Therefore, appropriate techniques should be incorporated in the planning, operation, and management of forest biomass-based production chains, to remove the forest biomass from harvesting locations and to mitigate the possible negative effects such as altering the flow of water bodies, reducing the productivity of forest soil, and destroying organisms that require deadwood for food and shelter purposes.

Loss of biodiversity, which can be defined as the extinction of species in a certain habitat, is highly related to environmental degradation. Biodiversity loss can be investigated in two categories, permanent losses of global species and changes in species composition. The adoption of biomass cropping systems to supply feedstocks to bio-energy and bio-products industries has the potential to alter the mix of biodiversity and ecosystem services realized from agricultural landscapes (Landis et al., 2017). Especially, land transformations that have taken place globally, to produce biomass-based products, have resulted in serious biodiversity declines (Joly et al., 2015). Therefore, the impacts of biomass-based production systems on biodiversity and ecosystem services should be well-assessed to determine the appropriate biomass types, plantation location, and agricultural practices to prevent the loss of biodiversity and consecutive reduction of ecosystem services.

4.1.3 Social Sustainability

Social sustainability can be defined as specifying and managing both positive and negative impacts of systems, processes, organizations, and activities on people and social life. The topics that social sustainability concept integrates include but are not limited to; health and social equity, human rights, labor rights, practices and decent working conditions, social responsibility and justice, community development and well-being, product responsibility, community resilience, and cultural competence. Although few conceptual frameworks have been developed, these changes cannot be quantified easily and accurately due to the complex and dynamic social, physiological, and anthropological factors inherent, and quantitative evaluation methodologies for such social perspectives is still immature. This may be one of the reasons that social sustainability issues have gained considerably less attention from researchers and decision-makers in comparison with economic and environmental sustainability issues, although these issues have a critical importance for the human life.

Biomass-based production chains might have a wide range of social impacts on the related regions, habitats, and residents. These impacts may pose positive or negative changes on human life, health, personal rights, well-being, and community, depending on the level of consideration of social sustainability issues in design, planning, and management procedures.

The main social performance indicators of a supply chain are, job creation and contributions to the employment pattern of the associated region, social and cultural acceptability, contribution to the rural development, and energy and food security. Some of the indicators such as job creation and employment affect as well as contribute to rural development, and can be seen as socioeconomic indicators, because they involve the interaction between the economic and social factors. Among all social sustainability indicators, job creation (and accordingly income creation) in biomass-based production chains is the most quantifiable one that is commonly used to measure the social impact of the chain. Job creation can be represented by quantity and quality of jobs created after the establishment of biomass-based production systems and supply chains. Strategic decisions about the size and configuration of the chain (i.e., the numbers, capacities, and locations of the production, storage, preprocessing, and distribution units), as well as the tactical decisions on the production and transportation levels and operational conditions made in design, planning, and management of the chain elements, specifies the quantity and quality of direct, indirect, and induced jobs created. Direct jobs can be described as the immediate or on-site jobs created, for example, jobs for constructing the production, storage, and other processing units, on-site contractors, and hired personnel for operational purposes are direct jobs. The establishment of the chain results in an indirect increase in economic activities that leads to indirect jobs, that is, the activities that occur when investors, producers, or contractors pay other suppliers in their business environment for necessary goods and services to be used in the production chain. Examples of this may include the suppliers of energy and other materials, bankers, and accountants who provide financial services. Induced jobs are the jobs induced by the increase of wealth and well-being, as a result of the increase in the number of the people who are directly or indirectly employed, in the elemental or in the business environment of the biomass-based production chain. The total employment effect raised by the establishment of the production chain units can be estimated by summing up the three types of effect. To systematically calculate and evaluate the social impact associated with job creation and to quantify the contributions to the local employment pattern of the region, both the yearly impacts from the permanent operations, such as production, distribution, and maintenance that continue along the lifetime of the units in the supply chain and the impacts from temporary activities, such as construction, should be taken into account.

Besides job creation, another important indicator of the social impact of biomass-based production chains is the contribution to the development of rural communities and its poverty alleviation potential. Contribution to rural development can be described as the contribution of the chain to the well-being and wealth of the rural population and residents in the environment of the chain. Such improvements can be achieved through supply of renewable energy and fuel for domestic and agricultural purposes and increasing the income per capita as a result of increased jobs. To measure and evaluate the potential social impact of the chain to the development of the rural community, it should be determined how the income generated from supply chain activities will be allocated and where it will be distributed and used (in the local community or somewhere else). Biomass-based production chains are typically constructed around rural areas that have the potential to produce agricultural biomass resources. These areas may be under the control of subsistence farmers with small-scale agricultural fields. Distribution of the income generated by the supply chain activities can create equity and improvement in the quality of life of nearby communities. A planned and scheduled development program along with a scheme that includes attractive biomass prices to farmers should be placed to gain the maximum benefit from such investments.

Social and cultural acceptability of biomass-based production systems, as well as the other types of renewable sources-based systems, are objects at issue in recent decades due to public perceptions concerning the esthetic, recreational, and cultural values of the local community. These concerns have a significant effect on the social sustainability of the production chain, because of the fact that a system can only be seen as a sustainable system on condition that it is acceptable by the local community. Social acceptability has a vital role in the existence, scale, and location decisions of biomass-based systems specifically in the design and planning phase. Social perceptions of biomass-based systems are mostly influenced by the issues that may be induced by the establishment of large-scale production systems and supply chains such as, land-use change for biomass production purposes and its potential indirect effects on the supply of food and animal fodder, loss of biodiversity, concerns about genetically modified energy crops, possible effects on human health and environment, and esthetics (including bad odor). Land-use change, biodiversity, and crop modification issues are generally encountered in plant-based biomass systems, whereas the latter two issues can be generally observed in waste biomass-based production chains (e.g., anaerobic digestion plants that convert waste biomass into energy and fuel). For the woody biomass-based systems (e.g., forest biomass and forestry residues), esthetics of the harvest areas, loss of biodiversity, and ecological restoration and wildfire risk might be included in this list. The other factors that influence the perception of local community and accordingly the social impact of the chain are, trust in the stakeholders, information-sharing and participation of the local community in the planning and design processes, effects on natural properties, and perceptions on community benefits. Furthermore, social acceptability of the system is also influenced by the possible effects of the system on social resources, such as water resources and the public areas that are utilized by the community. For this reason, location and distance of the biomass-based production, storage, preprocessing, and distribution plants should be planned by taking into account the distance to water bodies, wetlands and lakes, residential areas, natural reserves and protected areas, parks and recreational areas, coasts and touristic areas, natural heritage, archeological sites, and airports. Distance and location planning have critical importance in the establishment of large-scale systems. In addition, a reliable, clean, and sustainable system for water utility and treatment of wastewater from the supply chain should be constructed since clean and fresh water has vital importance for public health. Adapting appropriate systems scales, increased knowledge and awareness of public interest and the ecosystem, along with appropriate technical solutions in design, planning and management of biomass-based production chains, can enhance the social acceptability of the system.

Another social concern, namely food security, has been generated by the change of land for growing food crops into biomass production fields for energy, fuel, and other purposes, which has also resulted in food versus fuel debate. This issue is especially a matter for the production of bio-fuels such as bio-diesel and ethanol, which uses food crops (e.g., corn, sugarcane), grain, sugar, and oil crops as feedstock. Competition between food and fuel may result in increased food prices, as in the case of 2007–2008 global food crises (UN, 2011), however, several recent studies stated that bio-fuel production is responsible for a much smaller effect on food prices than initially expected (Dale et al., 2013; Trostle et al., 2011). There is a strong relationship between food security and social acceptability of the biomass-based production project in that the local community may object to the establishment of biomass plantations and may not want their arable lands to be used for feedstock production for energy, fuel, and other bio-product purposes. Food security can be measured by; (1) food price volatility, which depends on the food supply and demand in a region, (2) the level of land-use competition between food and fuel, which can be estimated by the proportion of food crop uses to the total amount of procured crops. The food security and sustainability of producing food crops should be considered in decision-making for the design, planning, and operation of biomass-based production chains.

4.1.4 Technical Sustainability

Besides three main pillars of sustainability (economic, environmental, and social), technical issues have to be addressed by decision-making procedures for wisely designed production units and network configurations along with logistic/transportation systems, and sustainable operations of biomass-based production chains. Technical sustainability can be defined as the technical and technological issues that need to be identified, for an efficient operation with reduced plant downtimes and uninterrupted product supply, increased productivity in conversion plants and increased operational efficiency in biomass cultivation, harvesting, storage, and treatment processes. The technical issues that can be encountered during operations are highly related to the maturity, reliability, and availability of the system and process technology. Production technologies, based on renewable sources, are relatively immature in comparison with the conventional nonrenewable-based production technologies. Hence, the technical sustainability issues should be handled in detail in the design and management of biomass-based production chains. Besides the reliability and maturity level of the chosen technology, the manageability of its daily operations, including the periodic maintenance and repair needs, is important for a technically sustainable system. For an enhanced and prompt maintenance and repair service, local availability of servicing needs and spare parts has critical importance. Once the issues related to technological maturity and manageability of biomass-based production systems are sorted out, an acceptable level of technical sustainability is achieved that increases decision-makers' and investors' confidence and willingness to invest in biomass-based production technologies.

Apart from the technological issues, some of the most important technical issues that affect the sustainability of the design of biomass-based production systems and chains are, the type, productivity and efficiency, and size/capacity of the conversion, storage, and preprocessing systems, which are interrelated so that the choice about one factor relates to the other. The type and quality of the final product and biomass sources specify the type of the technology required. The efficiency and productivity of the technology specify the quantity and quality of the final product per unit of biomass input and is associated with the type and scale of technology. The scale of the technology is important, in terms of economic viability of the system, and determines the quantity of product output as well. Establishment of larger capacity systems results in lower capital and operational cost per unit of production. However, the capital costs of biomass-based conversion systems are generally higher in comparison with conventional nonrenewable-based systems, hence it is essential to design the system so as to attain economies of scale. However, the practical capacity of the system is usually lower than its technical capacity, so that the amount of available biomass and the cost of biomass procurement and transfer to the system specify the operating scale. The type, scale, and productivity of the process technology significantly influence the capital and operational costs of the system. However, these factors impact the quality and quantity of the final product and hence the pricing strategies. The trade-offs between technical/technological issues and economic sustainability have to be considered while making decisions for design and management of biomass-based production chains.

There are also interactions between technical issues and the environmental sustainability of the system. The type and scale of the chosen production technology have a considerable role in environmental impacts, such as GHG emissions and soil contamination. Improper waste disposal and filtering practices may cause severe environmental issues around the production chain. The size of the investment also impacts the employment generation, that is, the number of jobs created, which is a social sustainability indicator of the system. The large-scale projects create more jobs than medium- and small-scale projects.

The technical design of the system has a significant effect on reducing the environmental problems, in other words, choosing the most suitable technology options for water treatment, waste management, emission control, and recycling can be helpful in minimizing the disposal and emission of waste into the atmosphere, soil, and water bodies. Also, using effective technologies, which use energy and fuel derived from renewable sources, for cultivation, harvesting, collection, treatment, and transportation of biomass and related bio-products, aids in mitigating the disadvantages of using fossil fuels.

4.2 DESIGN, PLANNING, AND IMPLEMENTATION OF SUSTAINABLE BIOMASS-BASED PRODUCTION CHAINS

4.2.1 Sustainability Indicators, Standards, and Certification Schemes

Biomass-based products and related by-products distinguish from nonrenewable source-based products by causing significantly less environmental pollution and eliminating the disadvantage of depletion and high prices of nonrenewable sources. However, not all biomass-based production applications result in the same benefits. Throughout the world, it is commonly observed that improper production and logistical practices (e.g., transportation and storage), may lead to undue socioeconomic and environmental impacts such as excess GHG emissions, especially in large-scale production facilities with new technological forms of bio-product promotion and long-distance transportation. In addition, biomass crop production and the removal of biomass residues from forest and agricultural systems for energy purposes may result in changing land-use patterns and land competition with local food crops. To address these environmental, economic, and social concerns properly and to ensure sustainability in production and other activities in biomass-based production and supply chains, verification of biomass-based sources and related products is essential. In the context of sustainable development, the use of biomass sources for the generation of a range of bio-products, such as bio-fuels, bio-energy and bio-chemicals, is only justifiable if biomass and bio-products are produced and distributed in environmentally, economically, and socially sustainable ways. With considerable developments and increase in biomass-based production, ensuring sustainability in operations throughout the chain is becoming a key concern and is currently being considered as an important requirement for market access. Especially in recent years, apart from ambitious targets related to quality and quantity of biomass-based products, there is a clear demand from stakeholders that the biomass procurement, bio-product production, and all transportation activities take place according to a minimum set of sustainability considerations and requirements.

Sustainability criteria can be defined as principles that cannot be used as a direct measure for sustainability performance but add meaning and operational capacity to sustainability principles and standards. Sustainability criteria can be seen as points that integrate information from sustainability indicators, and allows interpretation and assessment processes, depending on this information. Indicators represent quantitative or qualitative factors or variables of the system that provide a base for assessment of system performance on the scale of acceptable standards of performance and measurement of achievements. Indicators are needed to assess economic, environmental, and social sustainability of biomass-based production systems and identify and quantify the sustainability attributes of different biomass, technology, process, or product alternatives. Table 4.1 depicts the most commonly used sustainability indicators throughout the world along with corresponding descriptions and units.

Sustainability certification has become a popular and important instrument for private companies and governments to show their sustainability performance on varying sectors such as energy, agriculture, trade, food, and forestry. Numerous certification schemes have been developed to demonstrate the sustainability of a product, service, system, or process in terms of satisfying certain standards and sustainability criteria.

Setting standards and establishing certification systems, based on evaluation of essential criteria and indicators, are possible strategies that can help to ensure that bio-products are produced in a sustainable manner and within certain quality specifications. Quality of bio-products, as well as safety of production and logistical activities, can be guaranteed by certification, which is the process where an independent third party evaluates and assesses the quality and safety of products, as well as operational and management practices, including production and procurement activities, considering a set of predetermined standards and criteria. Briefly, certification schemes ensure that biomass or biomass-based products satisfy a certain sustainability standard. Criteria that have to be fulfilled for certification of a biomass source, bio-product, or production process should be operational and measurable.

A wide range of voluntary certification schemes have been introduced to ensure sustainability in production, treatment, and trade of biomass and biomass-based products in past decades. The core driver of these schemes is to stimulate the producers to comply with the regulations and to enable the end user to differentiate and select among different feedstock options, processes, and services by shedding light on their environmental and social characteristics. Certification schemes may have different scopes and assessment approaches-based on varying criteria and indicator sets and complexities. Some schemes exist at a local level, whereas others are governed internationally. In addition, they can be classified according to the biomass sources they cover, agricultural certification systems, forest certification systems, solid biomass for bio-energy certification, and general bio-fuel certification systems. Depending on the interests and priorities as well as the strictness, scopes, and the level of detail of standards, the selection and development of many certification systems vary between different parties, that is, private companies, institutions and associations, governmental units, and non-governmental organizations. The appropriate certification scheme that fits with the organizational and supply chain characteristics should be selected by the company or governmental unit. The selected scheme should also be credible, meet the legal requirements, facilitate the trade of the bio-product, fit with the end user needs, and it should be easy to comply with the requirements of standard.

TABLE 4.1 Sustainability Indicators, Brief Descriptions and Units

Economic Criteria	Brief Description	Indicator(s)	Unit
Profitability	A measure of economic viability of the production chain or system. Can be achieved by minimizing the unit production cost and/or maximizing the revenues from bio-product sales	• Return on investment (ROI) • Net present value (NPV)	Percent • National currencies
Life cycle costs	Total of the costs for each stage of the production chain that add value to the end product, e.g., feedstock procurement, preprocessing, storage, conversion, and transportation	• Total cost per unit of produced bio-product	• National currencies
Resource use	Amount of the resources used in the system or production chain that are critical for the continuity of the production process (especially nonrenewable resources)	• Depletion of nonrenewable resources • Fossil energy return on investment	• Tons/year • Ratio (MJ)
Energy security	Can be satisfied by adapting systems to fulfill the energy demand of the population reliably by quickly responding to energy shocks and mitigating long-term dependence on fuel and energy imports	• Fuel price volatility (Standard deviation of monthly percent price changes over one year)	• Percentage
Energy balance	A measure of the extent to which the biomass-based energy (e.g., bio-energy) displaces traditional nonrenewable-based energy (e.g., petroleum) in each stage of the production chain that add value to the end product	• Net energy balance	• MJ/year • MW/year
Productivity	Related to efficiencies of resource availability and use in biomass-based production, processing, storage, and distribution	• Processing efficiency of technology or feedstock • Unit amount of end product per year	• MJ/tons • Tons/ha per year • m^3/ha per year • MJ/ha per year • National currencies/MJ

(*Continued*)

TABLE 4.1 (Continued)

Economic Criteria	Brief Description	Indicator(s)	Unit
Environmental Criteria			
Life cycle environmental emissions	Life Cycle Assessment provides an estimate of all types of environmental emissions from biomass-based production and use, emitted by the production, storage, and processing of biomass source, conversion of biomass into bio-products, transport and distribution of feedstock and bio-product, storage, and treatment of bio-products and the end use of bio-product	• Life cycle GHG emissions • Emissions of non-GHG air pollutants, including air toxics (GBEP, 2011) • Disposal of liquid, semi-liquid and solid wastes	• Grams of CO_2 equivalent/MJ • mg/ha • mg/MJ • mg/m^3 • Percentage • Tons/MJ • Tons/m^3
Water-use, efficiency and quality	Related to the water availability, use efficiency and quality. It is an important measure for the amount of water required for biomass-based production that is met from available water resources, especially in regions with competing demands on surface or groundwater. It also measures the level of degradation of water quality	• The percentage of water taken from water sources for biomass-based production, to the total actual renewable water sources • Volume of water taken from water sources for biomass-based production, per unit of bio-product output • Percentage of pollutant loadings to water bodies resulting from biomass-based production to the pollutant loadings from total agricultural production	• Percentage • m^3/MJ • m^3/kWh • m^3/ha • m^3/tons • Percentage • kg of N, P and active ingredient per ha per year

(Continued)

TABLE 4.1 (Continued)

Economic Criteria	Brief Description	Indicator(s)	Unit
Biodiversity	Related to issues about biological diversity resulting from conversion of land important for biodiversity, cultivation of some species for biomass-based production, inappropriate agricultural, and forest management practices	• Area and/or percentage of land with high biodiversity or critical ecosystems converted for biomass-based production (GBEP, 2011) • Area and/or percentage of the land used for biomass- based production where invasive species are cultivated (GBEP, 2011) • Area and/or percentage of the land used for biomass- based production where nationally recognized conservation methods are applied (GBEP, 2011)	• Percentage • Hectares • km^2
Land-use and land-use change	Related to the change of the land originally used for agricultural purposes with biomass-based production environmental purposes, which may have adverse environmental (and social) impacts	• Total area of land used for biomass purposes compared to total national surface or agricultural land or managed forest area (GBEP, 2011) • Net annual rates of conversion between different land-use types caused directly by biomass-based production, e.g., conversion of arable land, natural forests, grasslands and wetlands, etc.	• Hectares • Percentages
Soil degradation	Related to the productive capacity of the land and soils, on which biomass-based feedstock is cultivated or harvested. Soil quality can be lowered by soil degradation as a result of climatic factors and poor agricultural practices	• Soil quality • Amount of degraded soil	• Percentage • Hectares • km^2

(*Continued*)

TABLE 4.1 (Continued)

Economic Criteria	Brief Description	Indicator(s)	Unit
Social			
Social acceptability	Integrates the factors related to public perceptions on the biomass-based production systems concerning the esthetic, recreational, and cultural values of the local community	• Opinion of local community • Transparency and reliability of the system • Stakeholder involvement • Complaint handling	• Number or percentage of supporting people • Number or percentage of timely and relevant performance reports • Number or percentage of documented responses addressing concerns and suggestions of stakeholder
Rural and social development	Related to contribution to the rural development in terms of contribution of the activities created in the chain to the well-being and wealth of the rural population and community in the environment of the chain	• Risk of catastrophe • Employment generation • Change in household income (i.e., wages paid for employment and net income from the sale of biomass sources and bio-products in the chain)	• Number or percentage of incidences of occupational injury, illness and fatalities • Probability of catastrophic event • Average number of work days lost to injury per worker • Number of new full time jobs created in the production chain • National currencies per household per year

(Continued)

TABLE 4.1 (Continued)

Economic Criteria	Brief Description	Indicator(s)	Unit
Food security	A measure of impacts of biomass-based production on the price and supply of a food basket at the national, regional, and/or household level	• Food price volatility and price inflation • Changes in production, demand, import and export for food, feed, and fiber	• National currencies • Percentage
Labor conditions	Related to developing and employing effective and fair labor standards on work and social policy for all activities throughout the chain	• Change in average unpaid time • Child labor • Payment and fair salary • Health and safety plans • Respect of human rights	• Hours per week • Number of child labor • Number or percentage of timely and relevant payments

There are three major elements of a biomass sustainability certification system; the sustainability standards and criteria, the chain of custody, and the management and audit system. In certification systems, varying standards and criteria may be used to evaluate and assess the biomass, bio-product, or production system in terms of their sustainability. The selection of suitable standards depends on their coverage of different sustainability aspects and the way they are measured. Often, the social sustainability standards focus on people, whereas the economic and environmental sustainability standards focus on profit and planet respectively. The most commonly used standards in certification procedures include; economic: cost-effectiveness, security of supply, competition for land and other natural resources; environmental: biodiversity, environmental pollution, GHG emissions, water-use and quality, and soil quality; social: labor rights and conditions, social well-being, competition with food, land rights.

Certified biomass or bio-products have to fulfill a number of the above-mentioned sustainability criteria from feedstock procurement to the end product. The chain of custody provides a mechanism for connecting and tracking sustainability information and claims for certified material from the feedstock procurement phase to the generation of the final product. The main purpose here is the implementation and verification of a control mechanism for each party in the production chain to ensure the traceability of the sustainably managed and certified material in the product or product line. The chain of custody can be implemented in four ways in the production chain, book and claim, mass balance, physical segregation, and identity preserved (NL Agency, 2012). Among them, book and claim has the lowest level of traceability of certified products, and there are no links between the physical flow of products and sustainability indicators. The product flows cannot be physically segregated from each other in this way. In the mass balance method, certified products can be mixed with noncertified products at any phase of the material flow in the production chain with the condition that the overall quantity is controlled by the company. Physical segregation ensures delivery of only certified products to the consumer, the certified products can only be mixed with other certified products, and mixing with other product flows is not allowed. The highest level of strictness and traceability exists in identity being preserved, which ensures that the certified products are supplied from identifiable sources, and cannot be mixed any other certified or noncertified product.

An important element of the biomass sustainability certification system is the management and audit system that integrates the principles to ensure a certain level of transparency and accessibility, complaints handling and management, involvement of stakeholders, and information-sharing with stakeholders. Specific procedures for planning and execution are also governed, as well as the type, frequency, and validity of audits, which should be clearly declared to the auditor for each case. Information, such as rights and responsibilities of certified companies, lists of certified companies, and summaries of assessment reports of companies, should be publicly available for a transparent, accessible, and reliable certification system. Also, during an audit, stakeholders should be given the opportunity to participate in the evaluation, or review the process to comment on the relevant issues, or just as observers.

The European Union has presented a number of sustainability criteria and mandatory renewable energy targets that have to be accommodated by companies to receive public support in production of biomass-based products (e.g., bio-fuels, bio-liquids, and bio-energy). One of the most important directives proposed by the European Commission, to provide legislative regulations for biomass-based energy production, with recognizing criteria and mandatory targets related to sustainability, is "The Renewable Energy Directive", which establishes an overall policy for the production and promotion of energy from renewable sources in the European Union. This directive sets a binding target of 20% final energy consumption from renewable sources by 2020. To achieve this, member countries have committed to reaching their own national renewables targets. There are a number of national regulations, acts, programs, and laws involving a wide range of sustainability requirements (e.g., limits for emission levels, technical efficiency, and productivity requirements), investment supports, subsidies, grants and promotions, conditions for eligibilities for green certificates, strategic plans, financial support schemes, and guidelines on authorization procedures throughout the world, for biomass production and biomass-based supply chains. The Green Electricity Act (Austria), the German Renewable Energies Act (Germany), the Investment aid programme (Germany), the Energy Investment Deduction Scheme (the Netherlands), the National Waste Management Plan (The Netherlands), the Renewables Obligation Order (the United Kingdom), and the Renewable Heat Incentive (the United Kingdom) are some of these regulations. For more information on national regulations on biomass and its uses, refer to Pelkmans et al. (2012).

To demonstrate and verify that their biomass-based production system and bio-products comply with the targets and sustainability criteria, especially those that are introduced by the Renewable Energy Directive (see the EU Renewable Energy Directive 2009/28/EC), a large number of companies have been involved in voluntary certification schemes introduced by the European Commission. Based on varying environmental, social, and economic standards, these schemes provide control mechanisms to check, for example, the land-use and land-change for biomass production (e.g., the land with high biodiversity and carbon stock was not converted for biomass cultivation), the level of GHG emissions saved by the production system, impacts to soil, water and air, compliance with the legal requirements, protection of labor rights, and community relations. Varying verification procedures may be adopted by the schemes, however, the changes that might be relevant to the Commission (e.g., the changes in auditing procedures) must be reported to the Commission, which controls and evaluates the changes, to determine whether the schemes still have the appropriate characteristics and cover the sustainability criteria adequately. The International Sustainability and Carbon Certification, the Round Table on Responsible Soy EU RED (RTRS EU RED), the Roundtable of Sustainable Bio-fuels EU RED, the Red Tractor Farm Assurance Combinable Crops & Sugar Beet Scheme (Red Tractor), and the Scottish Quality Farm Assured Combinable Crops scheme are some of the voluntary certification schemes recognized by the European Commission. Most of the schemes cover the entire biomass-based production chain including production, conversion, trade, and end use stages, except Good Agricultural Practices (GlobalGAP), Red Tractor, and the Sustainable Agriculture Network (SAN) schemes that focus on the first step of the biomass-based production chain, namely agricultural processes for biomass production. Biomass coverage scope of the schemes may vary as well, that is, a number of schemes are biomass specific (Bonsucro BSI – sugarcane, the Roundtable on Sustainable Palm Oil – palm oil, the RTRS – soy), whereas other sets of schemes have a larger scope but do not capture all types of biomass (Forest Stewardship Council – forestry biomass, Good Agricultural Practices (GlobalGAP), Red Tractor and the SAN – agricultural products, REDcert biomass produced in EU-27). The rest of the schemes involve all types of biomass. In addition, all schemes are global, meaning that they are able to certify biomass coming from all over the world, with the exception of REDcert and Red Tractor, which only accept biomass produced in the EU (NL Agency, 2011). For further information on voluntary certification schemes, refer to NL Agency (2011) and the web site of the European Commission (https://ec.europa.eu/energy/en/topics/renewable-energy/bio-fuels/voluntary-schemes).

4.2.2 Sustainability Assessment in Design and Management of Biomass-Based Production Chains

To assess the sustainability of existing biomass-based systems, supply chains and projects, and to evaluate the sustainability of different technical options for new investments, economic, environmental, and social analyses should be conducted in design, planning, and management processes. The purpose of sustainability assessment in supply chain design and management studies may be, investigating the feasibility and sustainability of proposed projects to provide baseline comparisons among different technology/process options, feedstock/product alternatives, locations in the establishment of a new system or supply chain, and investigating the feasibility and sustainability of proposed projects, comparing between the existing and new options in the pre-established supply chain, evaluating ongoing operations and activities in terms of sustainability aspects, to detect the out-of-control situations and take necessary precautions. Identifying the optimal design of biomass-based supply chains is not the main focus of sustainability assessment studies, however, they are utilized to provide the required information and analysis to evaluate, compare, and justify the design options related to a wide range of supply chain elements, or activities to be facilitated, in making specific decisions throughout the supply chain.

4.2.2.1 Economic Assessment

To appraise the economic sustainability, regarding the entire supply chain or its separate components, economic feasibility and performance can be evaluated employing economic assessment studies that comprise different methods and metrics. The commonly used metrics that can be used in the economic assessment process are described in Section 4.1.1. By carrying out economic assessment methods, the current/potential costs and current/anticipated benefits of existing/proposed alternatives for biomass-based supply chains, projects, or systems on a yearly basis, through their lifetime, are calculated to reveal the economic trade-offs between different technology, system, process, feedstock, products, policy, project, and configuration options, for the purpose of selecting the most viable alternatives. Efficient economic assessment studies provide a detailed and clear understanding of direct, indirect, and induced economic impacts of the handled system to propose an objective solution to the problem of identifying the most suitable options. The key issues for a successful economic analysis are, defining an accurate lifetime or duration for the project, system or supply chain to be assessed, identifying the proper discount rates to accurately evaluate the outputs, and defining the suitable parameters to measure and monetize the specific economic impacts of a wide range of costs and benefits related to different alternatives. The total value of alternatives should be evaluated by analyzing the current and future cash flows which then will be discounted to the present time according to given discount rates estimated based on the interest rates.

There are three main methods that have been used in a wide range of economic assessment studies for the purposes of; (1) monitoring the performance of the existing system related to economic sustainability, and (2) comparing several alternative options and identifying the most cost-effective option among them to attain particular objectives (e.g., producing predetermined quantity of products to meet a specified demand, achieving a targeted amount of GHG emissions, meeting a specified technological standard). These are, cost-effectiveness analysis (CEA), cost-benefit analysis (CBA), and multicriteria analysis (MCA). CEA takes into account the costs of the system/supply chain or alternatives that can be quantified in monetary terms. The results of CEA can be expressed as discounted cost per unit of production. The CEA involves ranking alternatives considering only their costs, hence it does not involve any benefits achieved by meeting the targeted objectives, which makes it a straightforward assessment method, in comparison with other economic assessment methods, by not capturing the impacts of changes in welfare as a result of implementing the selected investment or policy options. In order to consider both the costs and benefits that can be expressed in monetary terms, to reveal the economic performance of the system or to select between alternative options, CBA can be used. CBA calculates the ratio of benefits to costs or net benefits regarding each option, and selects the option with the highest ratio of benefits to costs or benefits surplus (difference between total revenues and total expenses). Due to several advantages such as taking into account both monetized costs and benefits, ease of use and adaptability to practical situations, CBA is the most commonly used economic assessment method for appraising and comparing options for biomass-based systems, supply chains, and projects. After identifying the alternatives to be compared, the CBA procedure starts with the identification, estimation, and quantification of the costs and benefits related to each option under consideration during the whole lifetime of the option, including the project development phase. As in the case of CEA, CBA requires all of the quantified costs and benefits that are expressed in monetary values and observed during the lifetime of the project, to be converted into present values by summing up their discounted versions. The results of CBA can be expressed using three metrics. The most commonly used metric to indicate the CBA results is net present value (NPV). The NPV of each alternative is calculated by subtracting the present value of costs from the present value of benefits, and the alternative with the largest positive NPV can be specified as the best option. The second way of identifying the results of CBA is using a benefit–cost ratio, which is the ratio of discounted total benefits to costs. The benefits regarding the options that have a benefit–cost ratio greater than 1 exceed the costs, and the option with the greatest value of benefit–cost ratio can be selected as the best alternative. Internal rate of return (IRR) is another metric that can be used to indicate CBA results. IRR is the discount rate at which the sum of the benefits of an option is equal to the sum of its costs. The options with IRR value, greater than the discount rate used in the analysis, can be considered among the good alternatives. The major drawback of using CEA and CBA in economic assessment of a biomass-based production chain is that both of them require identifying all costs and benefits in terms of monetary values. There is a range of economic valuation methods that can be used to determine the monetary value of costs and benefits regarding commercial products and services, however, the values of non-commercialized products and services, such as the ecosystem services and natural resources, cannot be easily quantified. A considerable limitation can be encountered to reliably estimate and quantify all costs and benefits in decision-making for the design and management of biomass-based production chains, in which these kinds of services and resources are utilized. Also, there may be some important environmental impacts that cannot be quantified accurately. In these cases, alternative assessment and ranking methods may be needed, instead of CEA and CBA.

The complex and multifunctional nature of renewable sources-based systems may require economic sustainability assessment methods that comprise multiple and variable criteria representing a range of costs and benefits derived from alternative options. Also, these criteria may not be quantified and expressed as monetary values, instead, they can be defined in terms of qualitative statements and other units. In these cases, MCA provides efficient and well-established methods for solving decision problems that integrate multiple and/or conflicting quantitative or qualitative criteria. For this purpose, a wide range of MCA methods have been employed to establish preferences based on comparisons and relationships between different criteria, which then will be used to compare and rank the alternatives. In MCA, it is critically important to define the criteria set and estimate the relative importance of each criterion relevant to each alternative accurately. MCA (in other words multicriteria decision-making), related methodologies as well as important economic, environmental, social and, technical criteria for decision-making in biomass-based production chains, will be covered in Chapter 6, "Basics of Decision-Making in Design and Management of Biomass-Based Production Chains." It should be noted that financial assessment is different from economic assessment, in terms of the type of revenues and expenses considered (e.g., incentives, subsidies. and taxes would not be considered in economic assessment).

4.2.2.2 Environmental Assessment

Environmental assessment can be defined as identifying, estimating, and evaluating the environmental impacts of existing and proposed projects, by conducting environmental studies, to mitigate the relevant negative effects prior to making decisions and commitments. Besides economic performance of the biomass-based production chain, various environmental loads and potential impacts of a feedstock, product, activity, process, system, or entire chain should be systematically assessed and evaluated. The environmental criteria should be involved in design and implementation of biomass-based production systems, because large-scale biomass production and trade can have negative potential environmental effects, due to increased transportation, storage, and production activities. Environmental assessment involves specifying and evaluating the environmental consequences of adapting a configuration design (in terms of number, capacity, and location of production, storage, preprocessing, and distribution plants for biomass resources and bio-products), an operational strategy (including cultivation, harvesting, collection, storage, preprocessing, and transportation plans for biomass resources and bio-products), a biomass feedstock alternative or a process/technology option (for conversion and preprocessing activities), prior to the decision-making. There is a wide range of areas, for which environmental assessment studies can be utilized for biomass-based systems including:

- Comparing environmental performance of different biomass feedstock for different purposes (e.g., heat and/or power generation, bio-fuel production, production of pellets, etc.),
- Comparing environmental performance of different conversion technologies and scales,
- Evaluating the environmental impacts of different operational strategies for biomass cultivation, harvesting, and collection,
- Evaluating the environmental impacts of different options for biomass and bio-product transportation vehicles, fuels, and operational strategies for logistics systems,
- Assessing the environmental performance of different bio-products; this assessment can be utilized in standardization and certification of biomass-based products,
- Evaluating the environmental impacts of different alternatives for storage of biomass sources and bio-products,
- Comparing the environmental performance of different biomass preprocessing options,
- Comparing the environmental impacts of different material options for construction of biomass-based production, storage, and preprocessing plants,
- Providing a perspective on resource depletion and environmental emissions to the organs of state for support in making policies, plans, and proposing legislations and incentive schemes about biomass-based production systems.

Life cycle assessment (LCA) is a powerful method for environmental assessment studies, which has been widely utilized to assess the environmental impacts of various biomass-based systems and production chains that generate bio-energy, bio-fuel, bio-chemicals, and other bio-products. The following subsection explains the LCA procedure in detail.

4.2.2.3 Social Assessment

Social assessment can be described as a set of studies to analyze, predict, evaluate, and manage the positive and negative consequences invoked by social changes, as a result of planned investments and projects, with the primary aim of ensuring sustainability and equitableness in the human environment. Social performance evaluation and impact assessment is typically employed for planned projects, however, the same procedures can be applied to appraise the social impacts of disasters, crises, as well as other natural and human-made catastrophes (i.e., epidemics, warfare, and mass migration) on humans.

Many social performance metrics have been introduced for social assessment of biomass-based production chains, systems, processes and products, however, most of them cannot be quantified, although quantification of the evaluation criteria is necessary for an objective appraisal. The type of social indicators that can be quantified, to be used in social assessment studies, are generally related to job creation and employment generation opportunities.

4.2.3 Life Cycle Assessment of Biomass-Based Production Systems

LCA is a successful method that has a wide range of application areas for systematic analysis and evaluation of the environmental impacts of different feedstock, products, processes, systems, activities, and supply chains from the point of origin to destination. LCA is an effective approach to analyze and quantify all emissions to the atmosphere, water bodies and soil, and to estimate their potential impacts on human health, global warming and climate change, ecosystem quality, and resources depletion.

The goal of the LCA studies may be quantifying and evaluating the environmental performance of an existing biomass-based system, or aiding strategic decisions in new investments, by comparing between different supply chain configuration alternatives.

The point of origin and destination covered in the study specifies the system boundary and depends on the scope and purpose of the study. The LCA methodology comprises a number of stages and procedures. To guarantee consistency and transparency in LCA applications, a number of regulatory standards and guidelines are presented by ISO (ISO14040:2006, 2006). According to these guidelines, four main phases constitute the LCA methodology, namely goal and scope definition, inventory analysis, impact assessment, and interpretation. Fig. 4.2 illustrates the phases of LCA.

In the first phase of LCA methodology, namely *description of the goal and scope of the study*, description of the system or project, the main purpose, scope, assumptions and limitations of the analysis, boundary of the system or geographical territory, and identification of the functional unit are defined completely. Depending on the purpose of the study, three main system boundaries are considered in LCA studies; cradle-to-grave, cradle-to-gate, or gate-to-gate. In a biomass-based production chain, cradle-to-grave LCA covers all supply chain elements and activities from the extraction of biomass resources (cradle) to the use and disposal of bio-products (grave). It incorporates two stages of the production chain, namely the upstream stage and the downstream stage, as described in Chapter 3 "Biomass-Based Supply Chains and Logistics Networks." The upstream stage includes the upstream of a production chain, from extraction of biomass sources to the main conversion process (e.g., biomass to fuel, energy, or other bio-products conversion), whereas the downstream stage covers the part of the chain from the

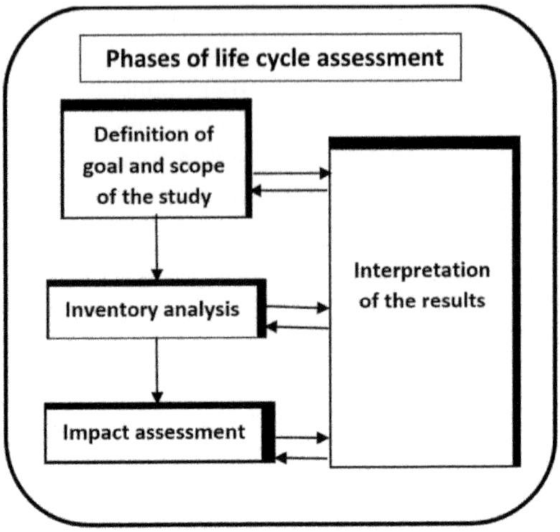

FIGURE 4.2 Phases of life cycle assessment.

conversion process to the product use and disposal phase. Cradle-to-gate LCA deals with all activities in a part of the supply chain, from resource extraction to the gate of the conversion, storage, preprocessing, or distribution plant, involving the life cycle stages from biomass cultivation, harvesting, collection, pretreatment, transportation, storage, and conversion. This type of system boundary is the most commonly adopted boundary, in the studies related to LCA, of biomass-based production chains. Cradle-to-gate LCA may either cover the processes in the upstream production chain or both upstream and downstream of the chain. Finally, gate-to-gate LCA covers a narrow part of the production chain, focusing on only one particular process in the entire chain. Apart from these commonly considered boundaries, there are three kinds of specific type of LCA that are used less frequently; cradle-to-cradle, well-to-wheel and ecology-based LCA. Cradle-to-cradle LCA handles the specific situations in cradle-to-grave assessment, where the used product is exposed to a recycling process instead of disposal, to minimize the environmental impacts of used products by a number of sustainable production activities. Well-to-wheel LCA takes place in the transportation sector, specifically used for transportation vehicles, fuel, and equipment. In a well-to-wheel LCA, the upstream stage comprises the production and treatment of fuel or energy for transportation purposes, and delivery of fuel/transmission of energy, whereas the downstream stage includes the transportation operation itself. Ecology-based LCA (eco-based LCA) focuses on the impacts of supply chain activities and services on natural resources and ecosystems, to manage these activities, in order to minimize the harmful effects. Besides specifying the boundary of the system to be captured by an LCA study, definition of the functional unit has critical importance for calculation and normalization processes. The functional unit is a reference unit, which is used to present the inventory and impact assessment results. The functional unit should be chosen so as to facilitate comparing the results of the LCA with the results belonging to reference products. In the case of biomass-based fuel production systems, this reference product may be a fossil fuel or an alternative bio-fuel product. For example, for a bio-fuel production system that produces biomass-based gasoline and diesel, "gasoline gallon equivalent", which indicates the amount of alternative fuel with the same energy content of one liquid gallon of gasoline, and allows consumers to compare the energy content of different fuels against a commonly known fuel – gasoline, can be selected as the functional unit for the LCA studies (Wikipedia, 2013).

The second stage of LCA is *inventory analysis*, in other words life cycle inventory (LCI), in which the data related to inputs and outputs of each process in the prespecified system boundary in a product's life cycle are gathered. To this aim, the data associated with inputs and outputs for a product, activity, or system in the biomass-based production chain related to biomass sources, material, natural resources, water, waste, energy, and environmental emissions are compiled, organized, quantified, and listed. The accuracy and transparency of the data, gathered in inventory analysis, have a critical impact on the efficiency of the following stages and reliability of the results. The list obtained by inventory analysis contains the quantities of a variety of atmospheric emissions, liquid, and solid wastes, biomass sources, materials, energy, water, and natural resources used in the production process, which can be categorized according to the stages of the life cycle, the medium for emissions (atmosphere, water, or soil), a product, or a process. LCI for a wide range of biomass-based products and processes can be obtained from various LCI databases such as, Ecoinvent (2008), the U.S. Life Cycle Inventory Database (2012), and the European Life Cycle Database (2006), and from commercial LCA software such as SimaPro and GaBi that use the Ecoinvent database. However, as the other renewable sources-based systems and technologies are relatively immature in comparison with the conventional nonrenewable sources-based systems, there may be still the need for data collection and documentation for new processes or products in biomass-based production chains. For this reason, in some cases, on-site measurements may be required to collect the related data.

Inventory analysis is a core element of LCA, however, inventory data only are insufficient for comparing multiple alternatives according to their environmental impacts and for selecting the preferable one(s) among them. It is also required to determine the severity and impact level of different environmental factors, of which data are collected in the inventory analysis stage. To this aim, the impact assessment stage can be included to further interpret the results of inventory analysis. The third stage of the LCA is the *impact assessment*, in other words life cycle impact assessment (LCIA), which involves the determination and categorization of the potential impacts of the handled system, activity, or product on the ecosystem, human health, and natural resource depletion. To this aim, the data collected, interpreted, and organized in the previous stage (inventory analysis stage), are used to understand and quantify the significance and magnitude of the potential impacts according to different environmental assessment categories. As a result, necessary actions can be initiated for prevention and remediation of the environmental burdens. A number of damage assessment models have been developed to calculate the numeric values regarding a range of environmental performance indicators, such as Eco-indicator 99 (Goedkoop and Spriensma, 2001) and IMPACT 2002 + (Jolliet et al., 2003), which are used by various commercial LCA software programs. Among them, Eco-indicator 99 and IMPACT 2002 + can be used to assess the environmental impacts in a broad range of categories such as human health, ecosystem quality, climate change, global warming, land occupation, and natural resources, whereas GWP concentrates on GHG emissions and their impact on global warming.

In the final phase of LCA, namely *interpretation*, the results of LCA are reviewed, analyzed, and discussed to provide reports involving suggestions and conclusions related to the systems, processes, activities, or products handled or compared. In this stage, the quantitative measurements related to different feedstock/product alternatives, process/technology options, configuration layouts, and operation strategies, obtained from impact assessment stage of LCA, are interpreted to compare the alternatives in terms of environmental sustainability.

The classical LCA methodology captures only environmental factors, and economic and social aspects cannot be handled by LCA. However, two life cycle-based methodologies, namely Life Cycle Cost Analysis (LCCA) and Social Life Cycle Assessment (SLCA) can be used to assess economic and social dimensions throughout an entire life cycle. LCCA is a method to analyze the total cost as well as a wide range of cost elements of a system, including design, development and production costs, investment costs, operation and maintenance costs, and disposal costs throughout its life cycle, to determine the cost-effectiveness of the options among different configuration, system, technology, process, service, feedstock, and product alternatives. SLCA can be used to analyze the social and sociological aspects of systems along the life cycle as well as their positive and negative impacts on humans and the environment. Quantitative, semi-quantitative, or qualitative techniques can be used for SLCA, in combination with LCA and LCCA, to capture economic, environmental, and social sustainability aspects simultaneously or on their own.

4.3 DISCUSSION

Sustainability of biomass-based systems varies extremely between feedstocks, technologies, and products and depends on many criteria that makes the assessment a significantly complicated and challenging task. Therefore, the selection, measurement, and accurate interpretation of the appropriate indicators and assessment methodology for the conditions of the handled system have a significant impact on the assessment performance, as well as identifications of the system boundary, reference unit/scenario, and other assumptions. However, even in a well-defined context, the decision-maker may face a wide range of uncertainties and sensitivities due to the lack of data and contradictions in objectives of different stakeholders.

Most sustainability concerns on biomass-based production are directed at the possible negative impacts of large-scale biomass-based production and trade, such as competition with food/feed production, in terms of land-use, natural resource depletion, soil contamination, and loss of biodiversity. Various biomass-based systems/chains are operated throughout the world, consisting of different biomass feedstock production systems, pretreatment and conversion operations, as well as transportation methods for raw materials. In most cases of biomass supply from production areas to production points, long-distance international transport can be necessary, which requires additional logistics with additional costs, energy consumption, and material losses, compared to local utilization. Certification is a useful tool to stimulate sustainability frameworks. However, development of interregional and international biomass/bio-product trade enforces addressing additional measurements and tools across multiple spatial scales at regional, national, and international levels.

Since sustainability is a critical issue that should be considered in design and management practices to prevent the negative economic, environmental, and social implications of biomass- based production chains, and to ensure the continuous utilization of biomass sources for generation of energy, fuel, chemicals, and other bio-products, different sustainability aspects have been covered by various studies on biomass-based production chains, which particularly cover the topics of sustainability assessment, design optimization, and planning of management and operational practices. However, environmental and economic sustainability of biomass-based production have been dwelt on in the large majority of these studies, and while there are fewer assessment and design studies that take into account social concerns, even they cover the social sustainability issue together with the other two dimensions of sustainability. Some studies involve quantifiable social sustainability indicators (mostly associated with employment), in conjunction with economic sustainability indicators, to reveal the socioeconomic impacts of the chain. Quantitative methods, to assess and evaluate the social sustainability of biomass-based production systems (e.g., impacts on community well-being, human health, welfare, and equity), are scarce and have still to be developed.

LCA is a well-known method to evaluate a wide range of environmental burdens and impacts throughout the stages of the life cycle of bio-products and to reveal the environmental trade-offs among alternative options. However, LCA results cannot provide comparative information on the overall environmental superiority of one alternative over the other one(s), because LCA results mostly depend on the subjective methodological choices of the decision-maker, such as the goal and scope definition, identification of the functional unit, and selection of models for impact assessment. Moreover, LCA lacks a systematic approach to capture various sources of uncertainty in LCIA. For a complete determination of the best alternative, LCA should be supported by additional methods and tools. By optimization methods such as mathematical programming, LCA results can be further interpreted, to obtain the most appropriate alternative, in terms of environmental impacts. To this aim, the life cycle optimization approach has been developed in recent years (Gebreslassie et al., 2013; Wang et al., 2013), which integrates optimization tools with traditional LCA methodology, both to assess the environmental impacts of various options for design and operation of the supply chain, and to determine the optimal alternative.

Furthermore, using multiobjective optimization in choosing the best alternative provides a wider view of sustainability aspects, capturing economic and social factors along with environmental impacts. Multiobjective optimization also has the ability to support more holistic comparisons in terms of sustainability perspectives, by analyzing the potential trade-offs between different objectives representing different aspects of sustainability. Modeling and optimization of biomass-based production chains, capturing different sustainability aspects simultaneously by employing LCA and multiobjective optimization approaches, are a promising area that have started to improve in recent years. Multiobjective optimization and related approaches are comprehensively explained in Chapter 7 “Modeling and Optimization Approaches in Design and Management of biomass-Based Production Chains.”

Due to their dynamic and complex nature, it may be hard to assess the sustainability of biomass-based production systems in terms of all three sustainability aspects. Furthermore, it may be challenging to set the most appropriate sustainability criteria to a biomass-based system, since there are numerous subcategories of each aspect, which may be contradictory and involve trade-offs. In these cases, multicriteria and multiobjective methodologies (Chapter 7: Modeling and Optimization Approaches in Design and Management of biomass-Based Production Chains, can be referred to for the details of these methodologies) can be used to determine the relative importance and preferences according to each subcriterion for the alternative options. In addition, a number of initiatives and certification schemes for biomass-based production systems cover many criteria.

To ensure sustainability in biomass-based production chains, appropriate certification schemes need to be applied, by both small producers and large conglomerates at global and local levels, considering the economic impact of certification procedures. To produce and trade a certified biomass-based product, some improvements have to be addressed in supply chain operations such as biomass production and supply, conversion, trade, and end use, which impose additional costs to the company. However, after the certification process, the increased credibility of the product may allow it to reflect this additional cost in the sale price of the product. Companies might choose one scheme or multiple schemes. Choosing one scheme, companies may face the disadvantage of not having sufficient supplies of certified biomass, however, choosing multiple schemes to overcome this drawback may result in higher costs. A better international coordination between certification initiatives is required, to preclude a fragmented biomass-based production environment with multiple national standards and quality considerations, and to improve the efficiency in further improvement of biomass certification schemes.

REFERENCES

Dale, V.H., Efroymson, R.A., Kline, K.L., 2013. Indicators for assessing socioeconomic sustainability of bio-energy systems: a short list of practical measures. Ecol. Indic. 26, 87–102.

Ecoinvent, 2008. The Ecoinvent database. <www.ecoinvent.org>.

European Life Cycle Database, 2006. European Commission Joint Research Centre <http://eplca.jrc.ec.europa.eu/ELCD3/>.

Gebreslassie, B.H., Slivinsky, M., Wang, B.L., You, F.Q., 2013. Life cycle optimization for sustainable design and operations of hydrocarbon biorefinery via fast pyrolysis, hydrotreating and hydrocracking. Comput. Chem. Eng. 50, 71–91.

Global Bio-energy Partners (GBEP), 2011. The global bio-energy partnership sustainability indicators for bio-energy, first ed. The Food and Agriculture Organization of the United Nations (FAO).

Goedkoop, M., Spriensma, R., 2001. The eco-indicator'99. A damage oriented method for life cycle impact assessment, Methodology report, third ed. PRé Consultants, Amersfoort.

ISO14040:2006, 2006. Environmental management – life cycle assessment – principles and framework. International Organization for Standardization.

Jolliet, O., et al., 2003. IMPACT 2002 + : A new life cycle impact assessment methodology. Int. J. Life Cycle Assess. 8 (6), 324–330.

Joly, C.A. et al., 2015. Bio-fuel impacts on biodiversity and ecosystem services. Chapter 16, Scientific Committee on Problems of the Environment (SCOPE) rapid assessment process on bio-energy and sustainability, 2014, 555–580.

Landis D.A. et al., 2017, Biomass and bio-fuel crop effects on biodiversity and ecosystem services in the North Central US, Biomass and Bio-energy. Available from: <http://dx.doi.org/10.1016/j.biombioe.2017.02.003>.

NL Agency, 2011. How to select a biomass certification scheme. NL Energy and Climate Change, Utrecht, The Netherlands.

NL Agency, 2012. Sustainability certification for biomass: Shaping the biomass market. Netherlands Programmes for Sustainable Biomass, Utrecht, The Netherlands.

Pelkmans, L., Devriendt, N., Junginger, M., et al., 2012. Benchmarking biomass sustainability criteria for energy purposes, Study carried out for the European Commission, Directorate-General for Energy 2011/TEM/R/190. Available from: <https://ec.europa.eu/energy/sites/ener/files/documents/2014_05_biobench_report.pdf>.

Tomlin, B., 2006. On the value of mitigation and contingency strategies for managing supply chain disruption risks. Manag. Sci. 52 (5), 639–657.

Trostle R., Mari D., Rosen S., Westcott P., 2011. Why have food commodity prices risen again? USDA Economic Research Service, Washington, DC, USA.

U.S. Life Cycle Inventory Database, 2012. National Renewable Energy Laboratory. Available from: <https://www.lcacommons.gov/nrel/search>.

UN, 2011. Chapter IV, The global food crises. Available from: <http://www.un.org/esa/socdev/rwss/docs/2011/chapter4.pdf>.

Wang, B., Gebreslassie, B.H., You, F., 2013. Sustainable design and synthesis of hydrocarbon biorefinery via gasification pathway: Integrated life cycle assessment and techno-economic analysis with multi-objective superstructure optimization. Comput. Chem. Eng. 52, 55–76.

Wikipedia, 2013. Gasoline gallon equivalent. Wikipedia, the free encyclopedia. Available from: <http://en.wikipedia.org/wiki/Gasoline> gallon equivalent.

FURTHER READING

Miao, Z., Shastri, Y., Grift, T.E., et al., 2012. Lignocellulosic biomass feedstock transportation alternatives, logistics, equipment configurations, and modeling. Biofuel, Bioprod. Bior. 6, 351–362.

Chapter 5

Uncertainty Issues in Biomass-Based Production Chains

Chapter Outline

5.1 **Uncertainty Sources in Biomass-Based Production Chains** **113**
- 5.1.1 Uncertainty in Market Demand and Prices 114
- 5.1.2 Uncertainty in Supply of Biomass and Other Materials 116
- 5.1.3 Technology and Process Uncertainties 117
- 5.1.4 Uncertainty in Governmental and Regulatory Policies 119
- 5.1.5 Uncertainty in Transportation and Logistics 119
- 5.1.6 Natural Uncertainties and Disruption Possibility 120

5.2 **Handling Uncertainties in Biomass-Based Production Chains** **121**
- 5.2.1 Approaches for Uncertainty Modeling 122
- 5.2.2 Methods for Uncertainty Impact Evaluation 137

5.3 **Discussion** **139**

References **141**

5.1 UNCERTAINTY SOURCES IN BIOMASS-BASED PRODUCTION CHAINS

Especially in recent years, a large number of related research and publications have pointed out the importance of making design and management decisions, considering a wide range of uncertainties that real-world supply chains face with, due to incompletely defined objectives and incomplete and/or unavailable information about the supply chain or its environment. Some of the serious problems that are caused by unaccommodated uncertainties in supply chains can be stated as follows:

- increase in the costs for supply chain activities, such as transportation and logistics costs, operational costs, investment costs, purchasing costs, etc.,
- excess response time to variations in the system, which slows down the response to changes in demands and decreases the customer satisfactions,
- bullwhip effect, as various linked facilities exist in the supply chain that effect each other's performance,
- difficulties in accurate inventory controlling.

Decision-Making for Biomass-Based Production Chains. DOI: https://doi.org/10.1016/B978-0-12-814278-3.00005-4

Biomass-based production chains have specific characteristics that make uncertainty consideration in design and management practices challenging, and more complicated, in comparison with typical nonrenewable sources-based production chains. These characteristics are mainly, seasonality of biomass sources and fluctuations in biomass production, due to natural conditions, such as soil quality, weather variability, pests and disease, perishability and degradation of biomass sources, and related supply shocks. Furthermore, biomass and bio-product markets are relatively young compared to other markets. These characteristics make biomass-based production chains more vulnerable and volatile than typical production chains, and are sensitive to fluctuations in the economic and financial factors. Furthermore, recent practices in international biomass-based supply chains, such as new marketing strategies due to the increased competition with nonrenewable-based systems and integrated design of large-scale systems, have increased the complexity and uncertainty in design and management of biomass-based production chains, thus making the application of decision-making strategies under uncertainty more critical.

Forecasts and estimations, based on historical data and past experiences, can be used to measure and quantify the uncertain parameters for effectively including them in the managerial decision-making process, and mitigate the possible negative consequences, caused by unexpected situations due to uncertainties. These forecasts may provide valuable recommendations for the adaptation to volatile market conditions and changing demands, fluctuations in natural conditions, price instability, and other factors that make the global competitiveness of biomass-based systems more challenging.

In designing biomass-based production chains, consideration of inherent uncertainties should be one of the major concerns, since uncertainty is the primary factor that has a significant influence in effectiveness of the design configuration and coordination of the system. Uncertainties may emerge from different phases and activities in biomass-based production chains. Some of the major uncertainties included by biomass to energy supply chains are about:

- market demand and price,
- supply of biomass and other materials,
- processes and technology,
- governmental and regulatory policies,
- transportation and logistics,
- natural conditions.

5.1.1 Uncertainty in Market Demand and Prices

Unknown or unpredictable variations in the quantity and timing of demand lead to demand uncertainty. Consumer demand can fluctuate due to reasons, such as seasonal patterns or changes in consumer preferences. For both biomass sources and bio-products, efficient demand management strategies incorporating estimation of the fluctuations in demand pattern is critical for mitigating the negative effects of demand instabilities on strategic decisions. Including demand and price uncertainties into the decision-making process can reduce expectation for profit generation; however, it provides more realistic information on expected profits.

Demand variation has a close relationship with local consumption behavior and regional economy structure. Demand uncertainty directly affects inventory policy as it involves the variation in sales during inventory replenishment. To protect against a stock out when demand exceeds forecast, safety stock should be added to the base inventory for both biomass sources and bio-products. Besides efficient management practices, the other factors to cope with the demand uncertainty are, availability of stable and reliable market conditions for both biomass sources and bio-products, stable and long-term governmental policies, including subsidies and incentive schemes, availability of stable market conditions for competitors, and substitute products. Reliability and flexibility of systems and processes for production/conversion, logistics and transportation operations are other critical factors. The flexibility of the system, technology, and operational practices are the most important factors that support the adaptability to changing demands. In addition, providing flexibility to the system, by introducing different alternatives for markets and production pathways, for a specific biomass feedstock or bio-product to allow required substitutions during temporary demand fluctuations, may be helpful to mitigate the undesired effects of localized or short-run demand spikes. The type of bio-product influences the flexibility of adaptation to changing demand patterns, that is, the drop in replacement products (e.g., bio-fuels that can be replaced with fossil fuels) has a strong advantage compared to incompatible products. Especially in the case of bio-fuels, the flexibility to changing demands increases with the availability of the required refueling infrastructure with the capability of fuel switching, and this infrastructure should also be provided to the end user if needed. In some cases, additional processes for refining, blending, changing compounds, or density of fuel may be required.

In the case of bio-energy generation, demand management is a key factor to ensure energy security, which has a strong relationship with economic security. Frequency of energy shocks, and dependency on external energy sources in a country, specifies the direction and level of the relationship between economic measures and energy security. Thus, the bio-energy demand from domestic, industrial, agricultural, and other activities should be estimated and managed by appropriate methods to prevent the energy shocks and long-term energy import. To enhance the energy security, sustainability and security of operations and resources in the biomass-based production chain, especially consistency in feedstock and water (and other auxiliary materials), supply and reliability of production operations should be improved, by taking necessary actions in case of highly variable commodity and service prices.

Price uncertainty can be defined as the chance or estimation that the price of a product might change. Price uncertainty may impact both costs and revenues of a production chain in either a positive or negative way; hence the severity and direction of fluctuation should be well-estimated to determine how the profitability of the chain will be affected by that change. As feedstock and product prices usually change within a small range due to competitive market conditions, price uncertainty in a supply chain is different from demand uncertainty, and can be predicted more accurately than demand uncertainty. In addition, fluctuations in the market price are more predictable than demand uncertainty, due to the long-term relationship between the procurement market and consumers. Price uncertainties in a biomass-based production chain typically include uncertainties about purchasing prices for biomass resources, costs of auxiliary material and utilities (e.g., water, steam, natural gas, etc.), bio-product prices, including power/heat price and costs related to different supply chain activities (e.g., transportation, storage, labor, construction, etc.). Generally, there is a strong correlation between the prices of nonrenewable sources and associated products (e.g., petroleum-based feedstock and products); however, the relationship between the prices of biomass sources and bio-products is more volatile because supply and pricing of biomass feedstock are more dependent on natural conditions such as season, weather, soil, and climate. Also, biomass-based systems are younger and more immature than conventional nonrenewable-based systems, which results in the lack of experienced labor. The integration of these two factors, namely fluctuations in biomass supply and inexperienced workers, may lead to some disturbances and errors in operations, which may further result in undesired process outcomes such as unqualified bio-products, delay in lead times and back orders.

5.1.2 Uncertainty in Supply of Biomass and Other Materials

Supply uncertainties in a biomass-based production chain are mainly characterized by three factors; variations in raw material yield, quality changes, and variations in supply lead time. The primary reasons of uncertainty about yield (or quantity) of biomass sources may be:

- shortage of feedstock,
- operational disturbances and improper cultivation of biomass crops,
- improper handling, storage or transportation practices that cause degradation on the biomass,
- seasonality of the supply,
- changes in harvesting technology,
- environmental and economic viability of alternative biomass options,

- unavailability of arable land,
- unreliable procurement sources possibly due to the market establishment.

The supply of biomass and other necessary materials may also be hampered by bad weather conditions or congestion at biomass source locations. Seasonality and annual variability of biomass supply is one of the major reasons for uncertainty in yields. In addition, plant-based biomass resources are required to be planted, cultivated, and harvested through a systematic growing cycle. Furthermore, the harvesting practices, frequency, and timing may affect the yields significantly. Thus, careful planning and scheduling practices should be developed to meet the all-year-round demand with seasonal production and to ensure the quantity and quality of the biomass supply. Furthermore, there is often a time limit for post-harvesting operations such as packing, handling, storage, and transportation, due to the perishability of biomass resources. Proper practices for post-harvesting operations should be developed and used to prevent the deterioration, especially of agricultural and waste biomass sources. The need to maintain a continuous biomass supply to the conversion facilities leads to an operational challenge to manage the biomass inventories.

A flexible supply system needs to be developed for the biomass producers and logistical systems to respond different markets and demand peaks in a timely manner. Yield fluctuations can be accommodated in the production chain that is supported by planned and interlinked transportation systems and infrastructure (especially for long-distance delivery), as well as flexible and reliable conversion technologies that can be adapted to diverse feedstock in different proportions. Feedstock supply systems may be more resilient, when the uncertainty is caused by the availability of feedstock and management practices, however, it may be more challenging to provide feedstock supply stability when faced with variations due to unexpected disturbances such as disasters or disease. In addition, the stability of supply of new and emerging biomass may be exposed to additional uncertainty due to immature production technologies and operational practices employed by inexperienced labor. For example, in the case of algae production, pond disturbances may occur in sudden fluctuations in temperature or pressure. In management of biomass-based production chains, holding safety stock, procurement from multiple suppliers for the same material, lead time increase, supply of substitute material, and the future purchase of raw materials, are some of the methods for overcoming the supply uncertainties. Also, in configuration design of the production chain, the distance between producers and suppliers should be determined, so as to allow the procurement of materials in a timely and cost-effective manner when needed.

5.1.3 Technology and Process Uncertainties

Technology and process uncertainties may result in the inability to generate the products with the planned quantity and within the predetermined quality specifications. These types of uncertainty may affect the continuity and quality of operations in biomass preprocessing, biomass to bio-product conversion, and bio-product treatment processes. Some of the process uncertainties are caused by delays in raw material procurement, which are closely related to logistics system uncertainties. Also, machine breakdowns and lead time specifications in the biomass-based conversion system have significant impacts on production and process efficiency. Scheduling maintenance and repair times properly, by effective maintenance planning techniques, can prevent the unexpected interruptions during the production. There may be additional uncertainty regarding the service level and lead time due to the batch production processes. On the spot contacting with the service suppliers is important to minimize the negative effects of possible breakdowns and interruptions.

Similar to supply uncertainties, process uncertainties directly affect inventory decisions. With a proper inventory policy, influences of process uncertainties on the production yield and timing can be minimized. Having limited inventory might cause backlogs and losses in profit in case of unforeseen disruptions in production. Besides machine breakdowns, interruptions engendered by inexperienced workers may also result in changes in the process pathways and related fluctuations in production yield. Holding safety stock, lead time increase by renewal agreements with customers, and outsourcing are some of the methods for coping with the process disruptions and uncertainties, and related fluctuations in production yield.

Technical/technological uncertainties, which are stimulated by rapid technological evolution, exist in a biomass-based production chain, when it is not clear that all system requirements can be implemented using the selected treatment and conversion technologies at desired performance levels. If the decision-maker has a strong understanding of the requirements and specifications of the technology, negative impacts of these uncertainties can be mitigated. In case of late or deficient clarification of the main system requirements, the existing technology may not support meeting of the needs, and additional technology modification or replacement costs may be incurred. Similarly, with a moderate or low understanding of the technology specifications, it may not be possible to fulfill the system requirements. The technical and technological uncertainties are higher in cases of new biomass-based system installations, since using new or modified technologies is generally exposed to high uncertainty.

The uncertainties stimulated by technology evolution are encountered in biomass-based system implementations, since biomass-based production technologies are relatively young and immature compared to traditional nonrenewable-based production technologies. To enhance the competitiveness with nonrenewable-based systems, improvements in biomass-based production technologies are needed in the near future to increase the efficiency of the conversion process (i.e., conversion rate) and energy efficiency. Technological developments may also lead to a considerable decrease in unit production and material handling costs. Enhanced service levels and service times for maintenance and repair activities may be other benefits of technological developments in biomass-based production systems. In the design and planning of biomass-based production chains, technology selection decisions should be made carefully, by determining whether to employ mature technologies with less improving space or nascent technologies with considerable improvement potential. This decision impacts the specification of a proper technique to choose the most appropriate conversion and treatment technology among multiple alternatives. Besides safety stock, lead time increase and outsourcing options, tight agreements with the service providers may be a method for coping with the technological uncertainties.

5.1.4 Uncertainty in Governmental and Regulatory Policies

These type of uncertainties occur when the future acts of government policy are vague. Governmental and regulatory policies, including tax and incentive schemes as well as other supporting policies, are core drivers of the biomass-based production industry, to provide and maintain the competitiveness of bio-products with their nonrenewable counterparts in terms of economic and environmental performance. There are two main reasons for this. Biomass-based production systems have relatively high investment and production costs and they have lower market acceptance at the current stage of development, because investors and customers may find conventional nonrenewable-based technologies and products more reliable and mature. However, governmental and regulatory policies may be amended or repealed after the installation of biomass-based production systems due to changes in political environment or economic conditions. Therefore, significant deviations may occur between the expected and realized profits. Anticipated changes in the tax rates and incentives influence the strategic and operational decisions in biomass-based supply chains. Some regions may be given priority in strategic development plans by the governments and locating the biomass-based production plants in that privileged regions may provide extra subsidies. Such subsidies have a considerable impact on the system profitability and payback period, hence giving preference to decisions related to the location of the system, as to where these privileged areas should be. Also, some extra subsidies may be provided considering the scale of the system, that is, establishment of small- to medium-scale systems may be supported by special financial incentive schemes. The possible future changes in governmental policies should be considered in the configuration design of biomass-based production chains.

5.1.5 Uncertainty in Transportation and Logistics

Transportation and logistics uncertainties may be engendered by lack of information and coordination between shipper and customer. Geographic and natural restrictions are other important factors that influence the availability of transportation lanes, and thus the stability and reliability of transportation and logistics systems. Unstable logistics operations may result in unsuccessful deliveries of both raw materials and finished products, thus increasing the lead time and cost of the transportation. Subsequent delays in fleet scheduling, backlog and stock out, increased freight cost, volatile supply of raw materials, and increased warehousing costs are among the other consequences of unsuccessful deliveries. Uncertainties in transportation and logistics operations may also result in the lack of optimized containers and associated increases in freight costs. In case of transporting perishable material such as plant-based or organic waste-based biomass, the undesired fluctuations in transportation distance and time may be the reason for material degradation, and cause a decrease in the biomass to bio-product conversion rates. For example, in bio-energy production, the unit production costs are highly related to the energy content of biomass feedstock, which might be exposed to significant losses due to material degradation. Distribution network design, schedules for logistics operations, and transportation mode should be optimized, according to the geographic conditions in the design phase, to mitigate the impacts of such uncertainties and to guarantee the effective movement of fresh products and services. In addition, development and application of efficient inventory management practices by holding optimized quantity of safety stock may be another option to minimize the unexpected consequences of transportation and logistics uncertainties.

5.1.6 Natural Uncertainties and Disruption Possibility

Along with the above-mentioned uncertainty sources, biomass-based production chains can be exposed to facility disruption due to man-made or natural catastrophic events such as floods, hurricane, earthquakes, terrorist attacks, plant explosions, and economic crises. These kinds of disruptions usually have a low likelihood of occurrence, but their impacts on the production chain are prominent, in that they may cause malfunctioning of the whole supply chain or a part of it. Most of these destructive disruptions cannot be foreseen, however, historical data or experiences can be utilized to estimate the probabilities on the occurrence of these events as well as the severity of the possible disruption. Depending on the magnitude, location and type of the natural and man-made disruptions, they may have a significant impact on supply chain parameters, such as supply and delivery time of materials, prices, costs, and demand. Furthermore, they may lead to failures and reduced capacities associated with facilities and transportation links, or these facilities and links may be even eliminated. The information and metrics regarding the probability and severity of the possible disasters can be incorporated into the decision-making procedure in the supply chain design phase to enhance the resilience and robustness of the chain, although usually, additional investment and operational costs may be endured.

Apart from catastrophes, another type of natural uncertainty is related to natural conditions. In a biomass-based production chain, the supply of biomass sources is highly related to biological and natural conditions, in that supply lead times are subject to uncertainties and cannot be altered due to nature. Because of the weather, climate and soil issues, supply levels, supply times, and product quality, may fluctuate significantly. The degree of uncertainty in the supply of wood residues is lower compared to the uncertainty in the supply of other crop residues, because they are grown over several years, which make them available all the year round, and hence less changeable due to climatic conditions. However, the yields may fluctuate according to seasons so that it might be troublesome to collect the wood residues during cold and snowy months.

5.2 HANDLING UNCERTAINTIES IN BIOMASS-BASED PRODUCTION CHAINS

In today's global and highly competitive markets, more flexible supply chains have to be established to mitigate the negative outcomes of a wide range of uncertainties, as described in previous sections. In most real-world flexible supply chain problems, model parameters that represent the elements of the problem, such as target values of objectives, supply and demand amounts, and prices are frequently imprecise because some information is incomplete and/or unavailable over the planning period. Conventional deterministic decision-making methods cannot solve all imprecise flexible supply chain problems, especially when there is lack of available information or lack of certainty in available information.

Due to many complex and variable factors that exist in all the echelons of the supply chain, the decision-making process in design and management of biomass-based production chains should incorporate the inherent uncertainty sources, to mitigate the undesired impacts that affect the economic, environmental, and social performance of the chain. This section discusses the methodologies to incorporate uncertainties in design, planning, and management of biomass-based production chains. The main purpose of this chapter is to make a comprehensive introduction to the methodologies that are commonly utilized, and that have the potential to be utilized, in decision-making for design and management problems in biomass-based production chains under uncertainty. In addition, the suitability and practicality of the methods for different problem characteristics associated with biomass-based production systems is discussed to provide a perspective to the readers in selecting the most appropriate methods for a variety of problems. Explaining the complete mathematical background and theory of the covered methods is not within the scope of this book, however, the major resources that include the complete background and theory of the methods will be referred to in the related parts for interested readers.

The choice of the most appropriate method for optimization problems in an uncertain environment depends on the type of the available data for uncertain parameters as well as their source of uncertainty. According to Beyer and Sendhoff (2007), in expressing uncertain parameters, mainly three formats can be used to distinguish between miscellaneous types of uncertainties, deterministic-based parameters, probabilistic parameters, and possibilistic parameters. Deterministic-based parameters define uncertainties by introducing domains defined as intervals or crisp sets. Probabilistic parameters can be specified by probability distributions that are usually derived using historical data indicating the likelihood of an event is plausible or realistic. Finally, possibilistic values represent uncertainties through fuzzy measures or sets that generalize real numbers in the way that it refers to a set of possible values instead of one single value. In this case, each possible value has its own weight between 0 and 1 indicating the possibility of its occurrence, which is called a membership function. A number of operations research methods can be applied, based on the above mentioned three formats, to represent uncertain parameters to aid in decision-making problems, including vagueness and ambiguity, about design and implementation of biomass-based production chains by manipulating the data. The approaches for incorporating uncertainties in design and management of biomass-based production chains can be investigated under two broad categories, approaches for modeling uncertainties and approaches for uncertainty impact evaluation. Fig. 5.1 classifies the approaches for incorporating uncertainties that have a wide range of application areas in the studies on design and management of biomass-based production.

5.2.1 Approaches for Uncertainty Modeling

The former category includes the modeling and optimization methods that capture the uncertainty in problem parameters, objectives, and decision variables and incorporates them into the decision-making process to solve the problem, so as to accommodate these uncertainties and minimize their negative influences on design and planning decisions. The most commonly used approaches in this category are stochastic programming, robust optimization, and fuzzy decision-making methods.

5.2.1.1 Stochastic Programming

Stochastic programming is a well-known and efficient approach to be used in decision-making environments, involving uncertain factors and circumstances represented by random parameters (or stochastic parameters), with an available probabilistic description in the form of probability measures, that is, probability distributions or densities. Stochastic programming is a suitable approach to incorporate the business, such as the usual uncertainties

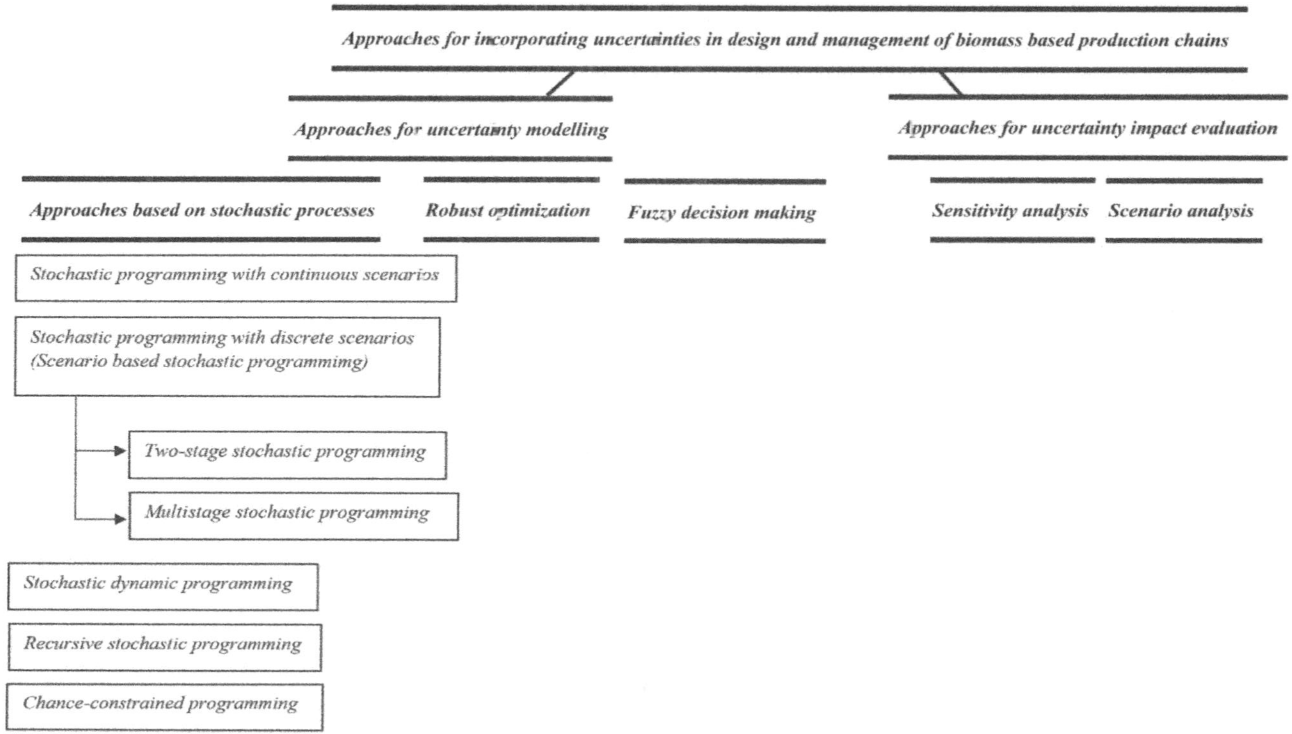

FIGURE 5.1 Approaches for incorporating uncertainties in design and management of biomass-based production chains.

related to availability of biomass resources, biomass and bio-product prices, economic and financial parameters, and variability in weather conditions.

The uncertainty in parameters can be identified by either discrete or continuous scenarios in stochastic optimization. Continuous scenarios represent the stochastic processes with parameters having a known continuous probability distribution, for example, the normal distribution with known mean and variance. The size of the model is maintained by this approach, however, nonlinearities and computational complexities may be introduced to the problem. There is limited research involving continuous stochastic parameters in the supply chain network design literature, and most of it handles the customers' demand and availability, or reliability of facilities, or transportation infrastructure as stochastic parameters.

Discrete scenarios are more frequently employed in supply chain design and management studies compared to continuous scenarios. *Scenario-based stochastic programming approaches* make use of discrete scenarios, through modeling stochastic parameters, using discrete scenarios with known probabilities. Using this approach, the computational complexity may substantially increase with the number of uncertain parameters, on the other hand, the computational disadvantage due to model nonlinearity may be eliminated. One of the key benefits of using discrete scenarios for modeling stochastic parameters is that they lead to more tractable models. In addition, this approach enables us to define various stochastic parameters by taking into account the dependency among them. Scenario-based stochastic programming approaches can be divided into two categories, two-stage stochastic programming and multistage stochastic programming (Birge & Louveaux, 2011).

Most of the problems related to problems in biomass-based production chains fit into *two-stage stochastic programming* due to the two-stage nature of decisions in supply chain design and management. In a two-stage program, the set of decision variables are divided into two main groups (Birge & Louveaux, 2011), first-stage decisions (or here-and-now decisions), are chosen before the realization of the random experiment and second-stage decisions (or recourse decisions or wait-and-see decisions), can be decided after the random experiment. The second-stage decisions can be stated as functions of the outcome of the random experiment and of the first-stage decisions. Once the random events are disclosed, further design or operational adjustments can be made through values of the second-stage or recourse variables. In other words, the values of the system variables are decided by the decision-maker in the first stage, and then the random event occurs, which might have undesired impacts on the decisions made in the first stage. In this case, a recourse action is taken in the second stage to compensate the negative consequences and infeasibilities of the first-stage decisions. In this procedure, the most effective results can be obtained, by defining a single policy in the first stage, and specifying a set of recourse decisions that form a decision rule based on different scenarios, stating the second-stage decisions that should be made in response to each random outcome. In case the second-stage decisions are continuous and have positive variables, a linear programming model can be solved for each scenario that represents the random outcomes, to calculate the value of recourse function for each feasible solution of first-stage decisions. However, if needed, non-linear equations could also be added in the objective function(s) and constraints, as well as integer and binary parameters/variables, to represent the decisions or problem specifications more realistically.

The general formulation of a two-stage stochastic programming problem is given by:

$$\begin{aligned} &Min_x c^T x + E_\xi[Q(x,\xi)] \\ &subject\ to\ \ Ax = b \\ &\qquad\qquad\qquad x \geq 0 \end{aligned}$$

$Q(x, \xi(w))$ is the optimal value of the second-stage problem denoted by below formulation:

$$\begin{aligned} &Min_x q(\xi)^T y \\ &subject\ to\ \ T(\xi)x + W(\xi)y = h \\ &\qquad\qquad\qquad y \geq 0 \end{aligned}$$

In the above formulation, x represents the vector of first-stage decision variables, and y represents the vector of second-stage decision variables. $\xi(q, h, T, W)$ includes the data of the second-stage problem with random elements. T and W are the matrices that denote technological and recourse data. A "here-and-now" decision x has to be made at the first stage, before the realization of the random vector ξ, of which probability distribution is known. At the second stage, after a realization of ξ, we optimize our behavior by solving an appropriate optimization problem.

At the first stage, the model optimizes the sum of the cost of the first-stage decision (denoted by $c^T x$), and the expected cost of the optimal decision is obtained in the second stage. Optimal recourse action is selected between the set of different possible recourse actions, which form the recourse problem in the second stage, through the costs assigned to recourse actions. Then, the expected cost of the optimal recourse action is added to the objective function. The second-stage problem can be considered as a penalty for the violation of the constraint $T(\xi)x = h$. In other words, the second-stage problem can be seen as an optimization problem that describes our supposedly optimal behavior when the uncertain data are disclosed, and its solution can be viewed as a recourse action with the cost denoted by $q(\xi)^T y$, where the term $W(\xi)y$ compensates for a possible inconsistency of the system $T(\xi)x = h$. The probability distribution of the random vector ξ can be reliably estimated from historical data, especially when the distribution does not significantly change over the considered time. Another way of deriving information on ξ can be through a simulation model with stochastic outputs.

The first- and second-stage decisions may contain sequences of decisions, events, and activities, thus the first-stage decisions influence the second-stage decisions. For example, strategic supply chain decisions, such as location and capacity decisions for conversion facilities should be made as first-stage decisions, before the realization of random parameters regarding operational and tactical decisions. However, when random parameters are disclosed, the operational and tactical decisions such as inventory, preprocessing, production, transportation, and routing have to be determined as second-stage decisions (Govindan et al., 2017). Depending on the assumption that stochastic parameters are distributed independently, two-stage stochastic programming is known to be nondeterministic polynomial time (NP)-hard. However, using some techniques, two-stage stochastic programming problems can be solved with a reasonable approximation. To this aim, decomposition methods are commonly used. One of the most widely applied decomposition techniques, for solving two-stage stochastic programs in biomass-based production chain problems, is Benders' decomposition (also known as L-shaped method). The progressive hedging algorithm, which is based on scenario decomposition, can also be used to obtain reasonable solutions. Another solution method is the branch and fix approach, which provides the coordination of the selection of the scenarios to be optimized jointly. For more detailed information on the techniques to solve two-stage stochastic programming models, refer to Birge and Louveaux (2011). Heuristic approaches are also widely utilized to obtain efficient solutions to NP-hard problems with a reasonable approximation in short central processing unit (CPU) times.

The second category in scenario-based stochastic programming is *multi-stage stochastic programming*, which can be used as a modeling approach in supply chain design and management problems with multiperiod settings and stochastic parameters. The decisions given through multistage stochastic programming cannot be based on outcome of random events in the future, in other words, a multistage stochastic program is non-anticipative. Instead, it can be formulated either as an iterative algorithm to calculate the total objective function through a recursive evaluation, by comprising a sequence of nested two-stage stochastic programs and including implicit settings to ensure non-anticipativity, or by explicitly imposing the non-anticipativity constraints. In multistage stochastic programming, the uncertainty associated with stochastic parameters are realized progressively in each of the multiple time periods composing the planning horizon considered in the problem, which also means that values of some decision variables have to be determined before realization of the random event, and some others are determined after the random event. It should be noted that, not all stochastic multiperiod supply chain design and planning problems necessarily require employing multistage stochastic programming.

In the literature related to biomass-based production chain design, planning, and management, multistage stochastic programming is rarely utilized, which means that there is a high potential in this field to develop and employ stochastic models with multiple periods using this method. Most of the stochastic decision problems in this field are modeled through two-stage stochastic programming, under the assumption that there is a single moment in the planning horizon for random parameters to become known, which means that their problem has a two-stage nature. However, in many real-world problems, the uncertainty and related randomness are realized progressively in more than one single moment in the life time of a project, which makes multistage stochastic programming a useful tool to represent the complex nature of biomass-based production chain problems.

In addition to developing multistage stochastic programs, proposing efficient approaches to obtain good solutions for them is another challenging issue. To solve multistage stochastic programs, nested decomposition procedures, Lagrangian relaxation-based methods (such as progressive hedging algorithm), and fix and relax coordination, which is an algorithmic metaheuristic (Albareda-Sambola et al., 2013), can be used. For more detailed information on the techniques to solve multistage stochastic programming models, refer to Birge and Louveaux (2011).

Scenario trees are widely utilized to solve multistage stochastic programs as an approximate representation of the multivariate stochastic processes, which have finitely many possible outcomes in each stage resulting from the discrete probability distributions for the random events (parameters). These outcomes are considered as scenarios, and each represents one possible disclosure of the uncertain future event, by fixing each random parameter to one of the possible outcomes. To form a scenario tree, scenarios can be obtained by sampling, resampling, and quantization methods. Starting from the initial scenario, the tree is established by bundling and deleting scenarios in a recursive way. This recursive procedure can work either backward or forward in time. All possible combinations of outcomes are enumerated to place all scenarios in a tree, of which branches form a path from the root of the tree to one of its leaves, representing each scenario. Each path visits nodes corresponding to assumptions of values of random parameters in the model.

An important aspect in scenario-based stochastic programming is the construction of an efficient set of scenarios that enhances modeling the underlying stochasticity problems. The most challenging part of generating an appropriate set of scenarios is that the parameters should be correlated, both with each other, and across the time units. Developing appropriate scenario generation procedures, to establish efficient scenarios for stochastic programs, has attracted the attention of researchers. To this aim, there is a significant focus on evaluating the quality and stability (in-sample and out-of-sample stabilities) of the scenario generation method. Kaut and Wallace (2007) can be referred to for more information about quality and stability measures for scenario generation procedures. Besides generating appropriate scenarios, it can also be problematic to define the associated probabilities in real-life problems. In addition, the sufficient number of scenarios may result in a large-scale optimization problem, with an excess amount of parameters and decision variables, which require reduction techniques. There are various efforts in stochastic programming literature to cope with scenario generation and reduction (e.g., Heitsch & Römisch, 2003; Høyland and Wallace, 2001).

Based on the type and linearity of the equations, variables, and parameters, the model may be Stochastic Mixed Integer Linear Programming, Integer Stochastic Programming, or Stochastic Mixed Integer Nonlinear Programming. Some stochastic approaches do not require a recourse action, by which the expected objective function is optimized, subject to a set of expected constraints. In this case, random parameters may be replaced by their expected values calculated, based on the average values. In a similar way to multistage stochastic programming, stochastic dynamic programming provides a powerful framework for addressing the dynamic behavior and uncertain operating environment of biomass-based production chains. *Stochastic dynamic programming* enables the inclusion of uncertain and risky parameters along with different time periods in the decision-making process. For example, problems about repeated or periodic failures in biomass conversion or pretreatment equipment as well as machinery replacement problems with stochastic failure, can be handled and modeled using stochastic dynamic programming. Planning of crop biomass cultivation and irrigation planning are examples of other periodic decisions that can be captured by stochastic dynamic programming. Like multistage stochastic programming, stochastic dynamic programming has been used in very few real-world implementations in the biomass production field, although they are promising methods for representing the dynamic linkages and periodic decisions in biomass-based production chains. Another variant of stochastic programming is *recursive stochastic programming*, which optimizes the system with stochastic parameters, by taking decisions sequentially through a discounted flow of returns, and then adjusting these decisions with additional information when available. It differs from stochastic dynamic programming in the way that decisions are fed back into the model.

Chance-constrained programming is another modeling approach depending on stochastic processes, which can be used when the uncertainties have to be expressed by qualitative measures, or when no recourse actions can be defined. Chance-constrained programming is commonly preferred in handling optimization problems with one or more constraints requiring to be satisfied with some probability or reliability level. By this approach, a specific probability is introduced to the objective or constraints of an optimization problem to be considered, as a chance factor. The final decision made, using chance-constrained programming, ensures a certain probability of complying with limitations defined by model constraints, in other words, a level of confidence is guaranteed as being feasible. So, utilizing chance-constrained programming enables us to define the economic performance of the supply chain, that is, the profitability, as a function of the prespecified confidence level of ensuring the system constraints, in other words, the reliability of the supply chain. This type of model can be solved through relaxation of the stochastic problem to obtain the nonlinear counterpart that can be handled by a commercial nonlinear programming solver. However, the calculation of the probability of satisfying inequality constraints poses the complexity of solving chance-constrained programming models. Chance-constrained programming models can be investigated in two classes, joint chance-constrained programming and separable chance-constrained programming. In the former, all soft constraints in the model are simultaneously satisfied with a given confidence level, whereas the latter takes each soft inequality separately and satisfies them with a high-enough probability.

5.2.1.2 Robust Optimization

Design and management problems can be optimized efficiently by optimization with a measure of robustness against the negative influences of uncertainties that are specified by a deterministic or set-based variability in the value of problem parameters or parameters of its solution. Robust optimization provides a feasible solution for any realization of the uncertainty in a given set for decision-making environments with incomplete or unknown information about the probability knowledge of uncertain phenomena. In this case, the uncertainty modeling may not necessarily be stochastic. The main purpose of robust optimization approach is for optimizing the worst case performance of the production chain, which is the most undesired realization of the uncertainty, and thus increasing the robustness of the production chain, which is treated as only a side effect in stochastic programming approaches. The worst case can be chosen either from a finite number of scenarios, for example, historical data, or continuous, convex uncertainty sets, such as polyhedrons or ellipsoids. Gabrel et al. (2014), can be referred to for more detailed information on robust optimization. The main advantage of robust optimization is its ability to capture the uncertain parameters, even when the actual information about these parameters is restricted, which is the case where stochastic programming approaches cannot handled in an efficient way. For this reason, rare disruptions in supply chains can be modeled more effectively by using robust optimization. Mulvey et al. (1995), defined two types of robustness, namely solution robustness and model robustness, meaning that the solution of a robust optimization problem is nearly optimal and nearly feasible in all possible realizations of uncertain parameters, respectively.

In robust optimization problems, the random parameters may be either continuous or discrete. Discrete uncertain parameters may be specified by scenario-based robust optimization programs, that is, discrete scenarios. Under this category, the robustness measures can be defined either with or without probability distributions. It is not possible to use the expected value criterion or other criteria based on probability knowledge in the case where the probability distributions of uncertain factors are not known. In this case, robust optimization can be applied by introducing appropriate robust measures for the problem. The most common measures in this class are minimax regret and minimax cost. The minimax regret measure obtains a solution minimizing the maximum relative or absolute regret, which is defined as the difference between the cost of a solution and the cost of the optimal solution for a scenario, whereas minimax cost is determined by minimizing the maximum cost for all scenarios.

For treating continuous uncertain parameters, these parameters are assumed to vary within some predefined intervals, in other words, uncertain data bounds. These intervals are also known as interval-uncertainties and this approach is called interval-uncertainty modeling. The key benefits of defining intervals are in protecting the optimization problems against deviations in uncertain parameters, which may lead to generation of unfeasible solutions and maintain computational tractability. To solve the optimization problem, multiple robust counterparts, which are deterministic equivalents of robust programs, can be formulated based on the structure of uncertain parameters. One of the key issues in constructing a robust version of the problem is maintaining computational tractability. Soyster (1973), defined the first step in interval-uncertainty modeling as ensuring that each feasible solution is also feasible for all disclosures of uncertain parameters within their prespecified intervals, by introducing a more tractable deterministic model as the counterpart of an uncertain optimization model. However, using this approach generally attains highly conservative solutions, which means that it may guarantee robust decisions to deal with the negative impact of uncertain parameters on the system performance, but may lead to losing optimality in solutions. After this study, several attempts have been made to eliminate the disadvantage of overconservatism. Among them, El Ghaoui and Lebret (1997), and Ben-Tal and Nemirovski (1998, 1999), developed approaches to generate less conservative solutions through nonlinear convex formulations, which are more difficult to solve and require more complex solution algorithms in comparison with Soyster's method. Afterwards, Bertsimas and Sim (2003, 2004) proposed a variety of robust optimization approaches that both provided an enhanced control of conservatism by using the idea of "budget of uncertainty" and resulted in a tractable linear programming model with computational simplicity, which can also be employed for optimization problems with discrete scenarios.

One shortcoming of all the above-mentioned robust optimization approaches is that all decision variables have to be determined before the occurrence of an uncertain event, which is not the case in most of the practical supply chain design and management problems with a multistage nature that require the determining of some decisions after disclosure of uncertainties. Addressing this challenge, Ben-Tal et al. (2004), proposed Affinely Adjustable Robust

Counterpart, a multistage robust optimization approach. This method enables us to make adjustable decisions that are affinely contingent on the primitive uncertainties. Another case in practical supply chain design and management problems, is that the distribution of uncertain parameters may be subject to uncertainty, and the moment that the information about this uncertainty of the distribution is available instead of the exact distributions itself. Distributionally Robust Optimization has been developed to cope with these situations by Scarf et al. (1958), which was further extended by Delage and Ye (2010), Goh and Sim (2010), and Wiesemann et al. (2014).

5.2.1.3 Fuzzy Decision-Making

Fuzzy decision-making environments provide appropriate methods to treat uncertainties in the form of ambiguity and vagueness. Ambiguity refers to the type of uncertainty in which the selection of multiple options among a set of alternatives is plausible. In other words, the meaning of ambiguous statements cannot be resolved definitely using a procedure consisting of a finite number of steps. However, a concept may be considered as vague if its extension is unclear or imprecise, due to the uncertainty about the objects that belong to the concept, or has difficulty in defining precise boundaries for some domains of interest. The major difference between ambiguity and vagueness is, that ambiguous concepts may be interpreted by specific or distinct statements, whereas it is nearly impossible to form any specific interpretations about vague concepts. Fuzzy decision-making approaches can be investigated in two classes, fuzzy flexible programming and fuzzy probabilistic programming. Fuzzy flexible programming models contain flexible goals and soft constraints that represent the flexibility of the target values of objective functions and the elasticity of constraints (Inuiguchi and Ramik, 2000). In other words, in a flexible programming problem, fuzzy goals and sets characterize the vagueness related to the decision maker's aspirations and constraints, respectively. Ambiguity in the coefficients of objective functions and constraints cannot be considered by fuzzy flexible programming, which is the key drawback of this approach. Fuzzy probabilistic programming makes it possible to deal with ambiguous coefficients in objective functions and constraints. When it is difficult to determine the exact value of model parameters, fuzzy probabilistic programming provides an efficient framework, by using probability distributions to model fuzzy parameters based on limited available data, as well as knowledge and experiences of decision-makers.

Fuzzy optimization is a method for dealing with the ambiguity and vagueness in uncertain parameters, represented by fuzzy elements, of which membership to a specific set is imprecise. One of the key benefits of fuzzy optimization is that it enables us to handle a wide range of uncertainties in the problem structure, such as the variability in the decision maker's aspiration level regarding the objective(s), the variability in the range of coefficients of objective function(s) and constraints, and the uncertainty in the satisfaction level of constraints, in contrast to the robust optimization approach, which captures the uncertainty in only parameters.

Due to the need for dealing with information arising from computational perception, that is, uncertain, imprecise, vague, ambiguous, or without sharp boundaries, firstly Zadeh (1965), developed a mathematical approach and named it Fuzzy Set Theory (FST). Since the acceptance of this theory, it has been applied to many fields in engineering and science. Some examples where FST has been successfully used are, control systems, knowledge-based systems, image processing, power engineering, industrial automation, robotics, consumer electronics, artificial intelligence/expert systems, management, and operations research. FST has been applied to many fields of operations research such as decision theory, project management, inventory control, scheduling, supply chain management, queuing theory, maintenance and reliability, and quality control. FST has been integrated with modeling and optimization techniques of linear and nonlinear programming, dynamic programming, integer programming, simulation, and multicriteria and multiobjective decision-making. Optimization and modeling approaches, integrated with FST, provide the appropriate framework to describe and treat inherent uncertainties, when there is a lack of evidence available or a lack of certainty in evidence, to cope with the imprecise nature of flexible supply chains.

FST: An element of a crisp set can either belong to the set or not in the conventional set theory, which means that it can either be 1 or 0. In FST/fuzzy logic, this condition was relaxed by Lotfi A. Zadeh, so an object has a degree of membership in a set, a number between 0 and 1. Ambiguous, qualitative, or subjective terms in the definition of a concept in natural language such as slightly, very, or quite are addressed by fuzzy logic. For instance, the answer to a question about whether a batch of products is defective or not is more flexible than a simple yes or no, and can be supported by a real number that displays the batch as 23% defective. This number is the degree of membership, which represents a subjective grade of satisfaction within a given range. In FST, classical bivalent sets that are assessed in binary terms are usually called crisp sets. Fuzzy logic is designed to mathematically indicate uncertainty and vagueness associated with notional activities of human beings such as thinking and reasoning. It has the required flexibility and robustness to provide tools to overcome real-world problems with uncertain intrinsic and parameters, which are in approximate values rather than exact. The following stages can be applied in developing a fuzzy system:

- Determine whether fuzzy modeling approaches are suitable for the problem. If the information about the system is identified approximately, or there is missing information about the problem, then the problem can be handled using FST.
- Define the values of problem parameters and their ranges.

- Identify a membership function utilizing the ranges for each parameter. The number of membership functions is determined by the decision-maker, based on the number of uncertain/imprecise parameters included in the problem.
- Develop the representation of the handled system using membership functions that show the satisfaction level of system parameters.

If the set X denotes a collection of objects defined by x, then a fuzzy set F in X can be formulated as a set of ordered pairs $F = \{(x, \mu_F(x)) | x \in X\}$ where, $\mu_F(x)$ is called the membership function for the fuzzy set F. The $\mu_F(x)$ assigns each element of X to a membership grade between 0 and 1 (included). If the value of the membership function is restricted either to 0 or 1, then F is reduced to crisp set. It means that crisp sets are special cases of fuzzy sets, i.e., crisp sets are subsets of fuzzy sets.

Membership functions: Depending on the problem specific characteristics, membership functions or possibility distributions are used to formulate fuzzy numbers. Lai and Hwang (1994), stated that, the grade of a membership function indicates a subjective degree of satisfaction within given tolerances. A membership function, usually denoted by "μ", associates each point in a fuzzy set F with a real number in the closed interval [0, 1]. It indicates the grade of membership of an element in a fuzzy set F. Thus, the nearer the value of membership function to unity, the higher the grade of membership of an element in a fuzzy set F. A membership function can be viewed as a quantification of the ambiguity of set F. Membership functions can take any form, but there are some common types that appear in real applications to represent the fuzzy set, including Gaussian, generalized bell, triangular, and trapezoidal membership functions. Among them, triangular and trapezoidal membership functions, which are explained in the following section, are commonly used in supply chain design and management studies.

Triangular membership function: A triangular fuzzy number can be stated as $\tilde{T} = (l, m, u)$ where, l is the lower limit, m is the expected value, and u is the upper limit of the number. If $l > 0$, then $\tilde{T}$ is called a positive triangular fuzzy number. The membership function of a positive triangular fuzzy number is represented by the following equation:

$$\mu_F(x; l, m, u) = \begin{cases} \dfrac{x-l}{m-l} & ; l < x \leq m \\ \dfrac{u-x}{u-m} & ; m < x \leq u \\ 0 & ; otherwise \end{cases}$$

Fig. 5.2 depicts a triangular membership function defined by $\mu_F(x; 20, 60, 80)$.

Trapezoidal membership function: A trapezoidal fuzzy number can be stated as $\tilde{T} = (l, m, n, u)$ where, l is the lower limit, $[m,n]$ is the range for

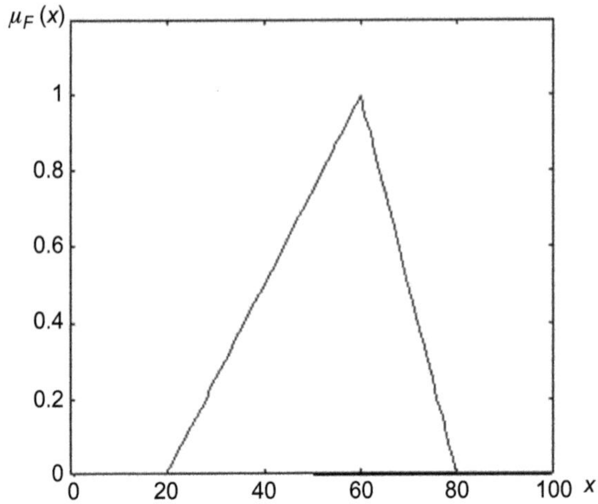

FIGURE 5.2 Triangular membership function defined by $\mu_F(x; 20, 60, 80)$.

the expected value, and u is the upper limit of the number. If $l > 0$, then $\tilde{T}$ is called a positive trapezoidal fuzzy number. The membership function of a positive trapezoidal fuzzy number is represented by the following equation:

$$\mu_F(x; l, m, n, u) = \begin{cases} \dfrac{x-l}{m-l} & ; l < x \leq m \\ 1 & ; m < x \leq n \\ \dfrac{u-x}{u-n} & ; n < x \leq u \\ 0 & ; otherwise \end{cases}$$

In Fig. 5.3, a trapezoidal membership function stated by $\mu_F(x; 10, 20, 60, 95)$ is depicted.

Fuzzy linear programming: In the case of parameters of the system to be optimized are fully defined and precise, the crisp Linear Programming approach (for detailed information on Linear Programming, Chapter 8 can be referred) can be used. But in practical supply chain problems, this assumption is unrealistic in that many imprecise or unknown parameter values are estimated by utilizing expert opinions. Thus, inclusion of FST in linear programming approach will be a good solution to these uncertainty/obscurity situations.

The concept of integration of decision-making with FST is proposed by Bellman and Zadeh (1970). After that, Tanaka et al. (1974), Zimmermann (1976), and Zimmermann (1978), further developed the concept of inclusion of FST into conventional linear programming problems. They dealt with linear programming problems with fuzzy goals and fuzzy constraints. Since

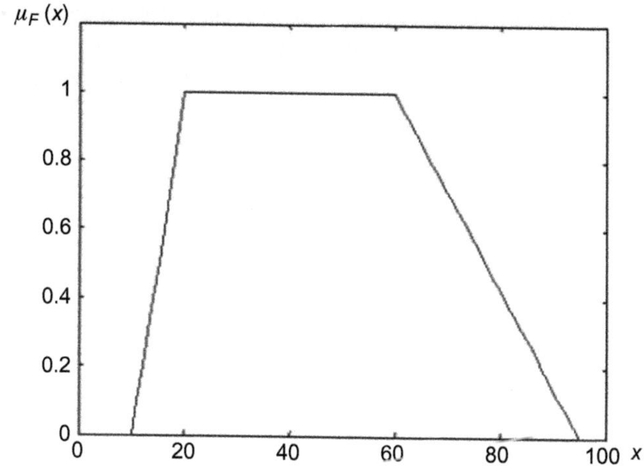

FIGURE 5.3 Trapezoidal membership function defined by $\mu_F(x; 10, 20, 60, 95)$.

then, a number of researchers have studied fuzzy linear programming problems (Tong, 1994; Gasimov and Yenilmez, 2002; Lai and Hwang, 1992). The general linear programming problem with minimization objective subject to the m linear inequality constraints can be stated as follows:

$$\begin{aligned}
&\textit{Minimize } Z = c_1x_1 + c_2x_2 + \ldots + c_nx_n \\
&\textit{subject to} \\
&\left.\begin{aligned}
&a_{11}x_1 + a_{12}x_2 + \ldots + a_{1n}x_n \leq b_1 \\
&a_{21}x_1 + a_{22}x_2 + \ldots + a_{2n}x_n \leq b_2 \\
&\ldots\ldots\ldots\ldots\ldots\ldots\ldots\ldots\ldots\ldots \\
&a_{m1}x_1 + a_{m2}x_2 + \ldots + a_{mn}x_n \leq b_m
\end{aligned}\right\} \\
&x_j \geq 0, \quad j = 1, 2, \ldots, n
\end{aligned}$$

where a_{ij}, b_i and c_j and are given parameters of objective function and constraints.

By introducing an n-dimensional row vector $c = (c_1, \ldots, c_n)$, an n-dimensional column vector $x = (x_1, \ldots, x_n)^T$, an m-dimensional column vector $b = (b_1, \ldots, b_n)^T$, and an $m \times n$ matrix $A = \lfloor a_{ij} \rfloor$ this problem can be defined in the following a vector matrix form.

$$\begin{aligned}
&\textit{Minimize } Z = cx \\
&\textit{subject to } Ax \leq b \\
&x \geq 0
\end{aligned}$$

The idea of softening the strict requirements of the decision-maker was firstly introduced by Zimmermann (1976), to consider that the gathered data about the problem can be uncertain or incomplete. Also, linear programming problems, to strictly minimize/maximize objective functions and strictly satisfy the constraints, may lead deceptive solutions. Zimmermann (1976), softened the general linear programming problem into the following fuzzy version (Sakawa, 1993):

$$\begin{aligned} & cx \prec z_0 \\ & Ax \prec b \\ & x \geq 0 \end{aligned}$$

where, z_0 is the decision-maker's aspiration level (target value of the objective function). The symbol $\prec$ denotes a fuzzified version of the ordinary inequality $\geq$ and can be read as "approximately less/greater than or equal to". These fuzzy inequalities representing the decision-maker's fuzzy goal and fuzzy constraints mean that "the objective function cx should be essentially smaller than or equal to an aspiration level z_0" and "the constraints Ax should be essentially smaller than or equal to b", respectively.

Zimmermann (1976), defined the general linear programming problem as follows considering the fuzzy goal and fuzzy constraints as equally important:

$$\begin{aligned} & Bx \prec b' \\ & x \geq 0 \end{aligned}$$

Where, $B = \begin{bmatrix} c \\ A \end{bmatrix}$ and $b' = \begin{bmatrix} z_0 \\ b \end{bmatrix}$.

Zimmermann (1976), also proposed the following linear membership function for treating the ith fuzzy inequality $(Bx)_i \prec b_i^{'}$, $i = 0, \ldots, m$, of the fuzzy inequality set $Bx \prec b'$.

$$\mu_i((Bx)_i) = \begin{cases} 1 & ;\ (Bx)_i \leq b_i^{'} \\ \dfrac{(b_i^{'} + d_i) - (Bx)_i}{d_i} & ;\ b_i^{'} < (Bx)_i \leq b_i^{'} + d_i \\ 0 & ;\ (Bx)_i > b_i^{'} + d_i \end{cases}$$

where, d_i states the limit of the allowed violation of the ith inequality. The value of the membership function varies linearly between 0 and 1. If the ith inequality is well-satisfied, the associated membership function will be 1. If the inequality is violated beyond the tolerance limit d_i, it will be 0. Fig. 5.4 illustrates such a linear membership function.

When the concept of linear membership functions is included in the Bellman and Zadeh (1970), fuzzy decision, the problem becomes choosing the x^*, which maximizes the minimum membership function values. This idea can be formulated with the following equation:

$$\mu_F(x^*) = \max_{x \geq 0} \min_{i = 0,\ldots,m} \{\mu_i((Bx)_i)\}$$

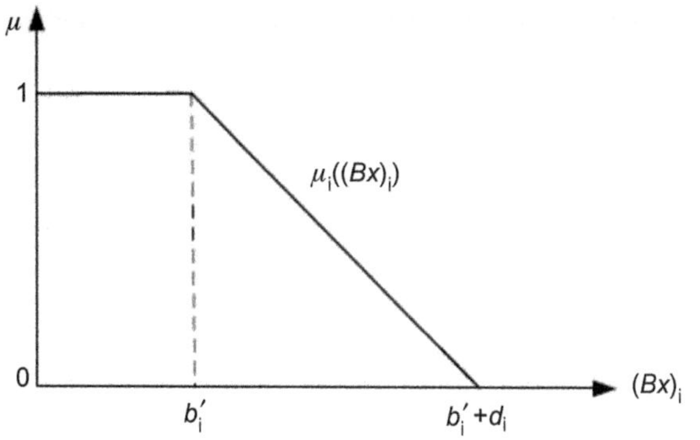

FIGURE 5.4 Linear membership function for the *i*th inequality.

As the aim is to maximize the minimum membership grade in the corresponding fuzzy model, Bellman and Zadeh's (1970), fuzzy decision is called the "min operator". By introducing the auxiliary variable λ, which is the overall satisfactory level of compromise, the following equivalent model can be obtained for the fuzzy linear programming problem:

$$\begin{aligned} &\textit{maximize} && \lambda \\ &\textit{subject to} && \mu_i((Bx)_i) \geq \lambda, \mathrm{i} = 1, \ldots, m \\ & && \lambda \geq 0, x \geq 0 \end{aligned}$$

5.2.2 Methods for Uncertainty Impact Evaluation

The methods in this category handle the parametric uncertainties, after making the design and planning decisions, so as to assess the impacts of changes in model parameters on the decisions and performance of the production chain. Two methods in this category are explained in this section, sensitivity analysis and scenario analysis.

5.2.2.1 Sensitivity Analysis

Sensitivity analysis is the study to measure the impacts of fluctuations in parameters of a mathematical model or system on the outputs or performance of the system. In other words, sensitivity analysis can be employed to apportion the changes in outputs of a system to different sources of uncertainty in its inputs. To this aim, one of the system parameters is changed by a certain percentage assuming all of the other parameters constant, the model is run and the percentage change of the prespecified performance indicator is observed. To obtain more reliable solutions, sensitivity analysis should be conducted in tandem with uncertainty analysis, which includes quantification and propagation of uncertainties.

Sensitivity analysis can be applied to explore the robustness and accuracy of the model results under uncertain conditions, and to comprehend the relationships between input parameters and performance indicators of a system or model, by revealing the unexpected relationships. Monitoring the impacts of variations in model parameters is useful in terms of the identification of the inputs that cause significant uncertainty in the performance indicators. Therefore, these significant parameters can be focused to reduce the uncertainty and increase the robustness and reliability of the system.

In design and modeling of biomass-based production chains, sensitivity analysis can be utilized to understand the contribution of a wide range of sources of uncertainty to the model output uncertainty and the economic, environmental, and social performance of the system. For example, the decision-maker may want to identify the supply chain activities that have significant impact on the level of total GHG emissions, or specify the most important cost drivers with a considerable effect on the overall profitability of the supply chain, to acquire better knowledge in order to make an informed decision. In addition, alternative supply chain configurations or system options can be compared considering a range of economic, environmental and social criteria corresponding to these alternatives, in terms of their impact on the system performance. Another field where sensitivity analysis may provide useful insights is in multicriteria/multiobjective decision-making. In these fields, the evaluation criteria or different objectives are associated with weights indicating relative importance of these criteria or objectives, that is, the larger the weight is, the more critical that criterion or objective is. Hence, determining the weights accurately is an important procedure for reliable solutions. To reveal the impact of changes as percentage in the weights on the solution, sensitivity analysis can be conducted. A small change in the weight of critical objectives and criteria will result in a significant change in the final results. Hence, a sensitivity analysis may shed light into issues about assigning weights not anticipated at the beginning of the study, and improve the efficiency of the results by allowing required corrections.

The major drawback of sensitivity analysis is that system variables are usually interdependent, so that changing one of the factors probably affects the other factors. For example, changing the bio-fuel sales price, to observe the impact of fluctuations in price levels on the economic performance of the system, may cause changes in sales volume as well. Also assigning an optimistic or pessimistic value to the parameter depends on subjective interpretation of the decision maker, which can adversely influence the objectivity and hence the accuracy and reliability of the analysis.

5.2.2.2 Scenario Analyses

Scenario analysis is conducted, to analyze the impacts of possible future events on the system performance by taking into account several alternative outcomes, i.e., scenarios, and to present different options for future development paths resulting in varying outcomes and corresponding implications.

Scenario analysis is the process of forecasting the expected value of a performance indicator, given a time period, occurrence of different situations, and related changes in the values of system parameters under an uncertain environment. Scenario analysis can be used to estimate the behavior of the system in response to an unexpected event, and may be utilized to explore the changes in system performance, in a theoretical best-case (optimistic) or worst-case (pessimistic) scenario. The occurrence probability and possible impact of a scenario should be considered in tandem to develop a strategic plan base on scenario analysis results. The major aim is, to analyze the results of the more extreme outcomes (with high probability and/or more severe impacts), to determine the investment strategy.

In design and management of biomass-based production chains, a decision-maker might use scenario analysis to estimate the impacts of several possible scenarios regarding changes in bio-fuel sale prices (increased, reduced, or constant prices), on the performance of the chain. Another analysis may depend on governmental strategies on incentivizing production from renewable sources or specific incentives for carbon sequestering operations. Different incentive policies may be considered to evaluate the behavior and performance of the production chain under changing financial and economic circumstances. The subsets of each of the possibilities and the correlations between these subsets may be taken into account, and the scenario weighted expected profitability of the production chain may be calculated.

The complexity of the problem, and the existence of stakeholders and related conflicting objectives in a supply chain, may make the scenario analysis a challenging practice. It may be difficult to forecast future events and corresponding impacts and assign probabilities to them. The system may need to be modeled by capturing possible fluctuations within a single scenario or possible correlations between multiple scenarios, which make the analysis further complex.

5.3 DISCUSSION

The main uncertainties that should be incorporated into the decision-making procedures in design and planning of biomass-based production chains can be investigated under two categories; (1) the inherent uncertainty in parameters associated with supply chain design and management parameters such as costs, prices, supply, and demand, and (2) uncertainties due to natural or man-made disruptions. The second type of uncertainty is less frequently encountered but has a more devastating impact on the pre-established facilities and transportation links in the biomass-based production chains compared to the first type of uncertainty. Hence, it is important to design reliable and resilient supply chain networks with a flexible procurement and distribution system, to reduce the potential impacts of disasters in the supply chain environment.

The impact of uncertainties on long-term strategic decisions, such as the establishment of a facility or infrastructure, are often more significantly compared to tactical and operational decisions, because the adaptation to future changes in strategic decisions and related implementations is more challenging and costly than changing weekly or monthly activities, such as distribution (including routing decisions) and storage of biomass and bio-products, in the supply chain. Hence, more comprehensive analyses should be conducted to incorporate the uncertainties when dealing with strategic decisions. Making design and management decisions for supply chains with highly perishable biomass sources and bio-products generally depend on the life and characteristics of the material, such as the rate of material degradation and associated loss in conversion potential. To mitigate the negative effects of uncertainties in the timing and flow of tactical and operational activities such as transportation and storage on the degradation rate of this type of material, this aspect should be handled in design, planning, and management practices. Defining long-term periods may provide the opportunity of changing strategic decisions, for example, changing the location and capacity of facilities over long-term periods, due to future needs. However, tactical or operational periods can cover the fluctuations in parameters in tactical and operational decisions, and enhance the flexibility of the supply chain due to a volatile and competitive business environment. Developing and employing comprehensive models and appropriate methods that capture multiple time periods is an effective way to capture the uncertainties and minimize the possible negative effects of them. The number of periods and the duration of each period should be specified considering the problem specific characteristics and the type of decisions that are more frequently considered during the entire planning horizon.

The selection of the most suitable methodology that enables us to incorporate uncertain parameters in the decision-making process also depends on the type and amount of available data about the problem. For example, if enough historical data can be collected to estimate the probability distributions of uncertain parameters, stochastic modeling may be the most appropriate tool. However, in many real-world problems related to biomass-based production systems, sufficient historical data cannot be gathered or the collected data may not fit into a distribution pattern, due to relatively young and immature technologies and processes, as well as the complexity about the procurement of biomass sources based on seasonality and dependency on natural conditions. Robust optimization may be convenient to handle such a situation. An emerging topic and recent trend in the modeling and design of biomass-based production chains under uncertainty is in developing fuzzy probabilistic programming models. Fuzzy probabilistic programming can be used to solve optimization problems with fuzzy random variables, probabilistic constraints with fuzziness, and a mix of deterministic, stochastic and fuzzy parameters. Fuzzy probabilistic programming covers fuzziness and randomness in problem parameters, objectives, and variables under one roof, and aims at getting the highest degree of satisfaction in achieving the optimal solution in a probabilistic environment. Apart from stochastic programming, robust optimization and fuzzy decision-making, sensitivity and scenario analyses provide frameworks to handle a wide range of uncertainties in design and planning problems. Sensitivity and scenario analyses can be conducted to capture uncertainties after the design and modeling phase, to determine the impact of different values of parameters and varying situations, on the economic, environmental, technical, and social performance of the system. These methods cannot be used to determine the most appropriate design options and management decisions under an uncertain environment, but instead they specify the significant parameters and their impacts on system behavior as well as variations in system conditions.

REFERENCES

Albareda-Sambola, M., Alonso-Ayuso, A., Escudero, L.F., et al., 2013. Fix-and-relax coordination for a multi-period location–allocation problem under uncertainty. Comput. Oper. Res. 40 (12), 2878–2892.

Bellman, R.E., Zadeh, L.A., 1970. Decision-making in a fuzzy environment. Manag. Sci. 17, 141–164.

Ben-Tal, A., Nemirovski, A., 1998. Robust convex optimization. Math. Oper. Res. 23 (4), 769–805.

Ben-Tal, A., Nemirovski, A., 1999. Robust solutions of uncertain linear programs. Oper. Res. Lett. 25 (1), 1–13.

Ben-Tal, A., Goryashko, A., Guslitzer, E., Nemirovski, A., 2004. Adjustable robust solutions of uncertain linear programs. Math. Prog. 99 (2), 351–376.

Bertsimas, D., Sim, M., 2003. Robust discrete optimization and network flows. Math. Prog. Series B 98, 48–71.

Bertsimas, D., Sim, M., 2004. The price of robustness. Oper. Res. 52, 35–53.

Beyer, H.-G., Sendhoff, B., 2007. Robust optimization—a comprehensive survey. Comput. Meth. Appl. Mech. Eng. 196 (33-34), 3190–3218.

Birge, J.R., Louveaux, F., 2011. Introduction to Stochastic Programming, second ed. Springer, New York, NY.

Delage, E., Ye, Y., 2010. Distributionally robust optimization under moment uncertainty with application to data-driven problems. Oper. Res. 58 (3), 595–612.

El Ghaoui, L., Lebret, H., 1997. Robust solutions to least-squares problems with uncertain data. SIAM J. Matrix Anal. Appl. 18 (4), 1035–1064.

Gabrel, V., Murat, C., Thiele, A., 2014. Recent advances in robust optimization: an overview. Eur. J. Oper. Res. 235 (3), 471–483.

Gasimov, R.N., Yenilmez, K., 2002. Duality in fuzzy linear programming. Algebras, Groups Geometries 19 (1), 39–56.

Goh, J., Sim, M., 2010. Distributionally robust optimization and its tractable approximations. Oper. Res. 58 (4), 902–917.

Govindan, K., Fattahi, M., Keyvanshokooh, E., 2017. Supply chain network design under uncertainty: a comprehensive review and future research directions. Eur. J. Oper. Res. 263, 108–141.

Heitsch, H., Römisch, W., 2003. Scenario reduction algorithms in stochastic programming. Comput. Optim. Appl. 24 (2–3), 187–206.

Høyland, K., Wallace, S.W., 2001. Generating scenario trees for multistage decision problems. Manag. Sci. 47 (2), 295–307.

Inuiguchi, M., Ramik, J., 2000. Possibilistic linear programming: a brief review of fuzzy mathematical programming and a comparison with stochastic programming in portfolio selection problem. Fuzzy Set Syst. 111 (1), 3–28.

Kaut, M., Wallace, S.W., 2007. Evaluation of scenario-generation methods for stochastic programming. Pac. J. Optim. 3 (2), 257–271.

Lai, Y.J., Hwang, C.L., 1992. Possibilistic linear programming for managing interest rate risk. Fuzzy Set Syst. 49, 121–133.

Lai, Y.J., Hwang, C.L., 1994. Fuzzy Multiple Objective Decision-Making: Methods and Applications. Springer, New York, NY.

Mulvey, J.M., Vanderbei, R.J., Zenios, S.A., 1995. Robust optimization of large-scale systems. Oper. Res. 43, 264–281.

Sakawa, M., 1993. Fuzzy Sets and Interactive Multi-objective Optimization. Plenum Press, New York, NY.

Scarf, H., Arrow, K.S., Karlin, S., Scarf, H.E., 1958. A min-max solution of an inventory problem. Studies in the Mathematical Theory of Inventory and Production. Stanford University Press, Stanford, CA, pp. 201–209.

Soyster, A.L., 1973. Technical note—convex programming with set-inclusive constraints and applications to inexact linear programming. Oper. Res. 21 (5), 1154–1157.

Tanaka, H., Okuda, T., Asai, K., 1974. On fuzzy mathematical programming. J. Cybern. 3, 37–46.

Tong, S., 1994. Interval number and fuzzy number linear programmings. Fuzzy Set. Syst. 66 (3), 301–306.

Wiesemann, W., Kuhn, D., Sim, M., 2014. Distributionally robust convex optimization. Oper. Res. 62 (6), 1358–1376.

Zadeh, L.A., 1965. Fuzzy sets. Infor. Control 8, 338–353.

Zimmermann, H.J., 1976. Description and optimization of fuzzy systems. Int. J. Gen. Syst. 2, 209–215.

Zimmermann, H.J., 1978. Fuzzy programming and linear programming with several objective function. Fuzzy Set. Syst. 1, 45–55.

Chapter 6

Basics of Decision-Making in Design and Management of Biomass-Based Production Chains

Chapter Outline

6.1 Decision Levels in Design and Management of Biomass-Based Production Chains 143
6.1.1 Strategic Decisions 144
6.1.2 Tactical Decisions 149
6.1.3 Operational Decisions 151
6.2 Decision-Making in Design and Management of Biomass-Based Production Chains 153
6.2.1 The Main Objectives and Criteria 154
6.2.2 Multiple Attribute and Multiple Criteria Decision-Making Methods in Biomass-Based Production Chains 165
6.3 Discussion 178
References 179
Further Reading 183

6.1 DECISION LEVELS IN DESIGN AND MANAGEMENT OF BIOMASS-BASED PRODUCTION CHAINS

In a significant number of studies in supply chain design and management literature, single-period modeling has been used, by which the decisions have been made for only one period. Although single-period modeling may prevent computational complexity, it may not be enough to obtain a responsive and resilient supply chain configuration and flexible decisions regarding to repetitive supply chain activities. The main reason is that through single-period modeling the time dependent fluctuations and variabilities in supply chain parameters may not be considered sufficiently while making decisions. Another part of the related literature has pointed out the supply chain design and planning problems with a planning horizon comprising multiple time periods. The periods considered in these studies can be classified into, strategic time periods and tactical/operational time periods. Strategic periods are considered in long-term decision problems usually about configuration design, which may need revisions in at least three years based on the

Decision-Making for Biomass-Based Production Chains. DOI: https://doi.org/10.1016/B978-0-12-814278-3.00006-6

business entity. Tactical/operational periods are used to model short- to medium-term decisions that are taken daily, weekly, or monthly, usually up to one year. These decisions generally refer to repetitive activities in the supply chain, such as distribution activities, demand management, changing inventory levels, and production scheduling. Tactical decisions usually aim to plan the mid-term activities, so as to provide profitable operations considering the limitations of the strategic decisions, whereas operational decisions help to achieve tactical goals, comprising detailed scheduling and planning for one day or week of operation. The time span covered by these three decision levels is mostly dependent on the nature and characteristics of the supply chain and related activities and it can vary for different supply chain structures. The design and management of efficient supply chains in today's competitive environment should focus on optimizing all the decisions to achieve a robust and reliable supply chain with a holistic management approach. Fig. 6.1 illustrates the hierarchy of decision-making for design, planning, and management of biomass-based production chains.

6.1.1 Strategic Decisions

Strategic decisions have a long-term impact on the performance of the supply chain which may need revisions after a long time period of usually three or more years. It focuses on the structural decisions related to the chain's configuration and infrastructures, the strategies to allocate resources, and the technologies to perform processes at each stage. They are usually taken at the highest levels of management, include a wide range of uncertainties and carry higher levels of risk. However, effective strategic decisions lead to increased profitability and high levels of rewards for all stakeholders in the supply chain. The issues that require strategic decision-making in biomass-based production chains include:

- choosing the location, site, installed capacity and purpose of biomass-based production and storage (centralized or decentralized) facilities,
- selection of conversion and treatment technologies and processes,
- configuration of a network of suppliers, distributers, and logistics handlers,
- long-term inventory management strategies,
- long-term technological improvement and capacity planning analyses,
- sourcing strategies, long-term supply, and demand contracts,
- ensuring economic, environmental, technical, and social sustainability.

Once made, strategic decisions cannot be altered easily in short terms and their influences can be observed over years and decades, even beyond the lifetime of the system, due to the high capital investment and set-up costs around the construction of facilities and technologies. As stated previously, the most important strategic decision is the configuration design

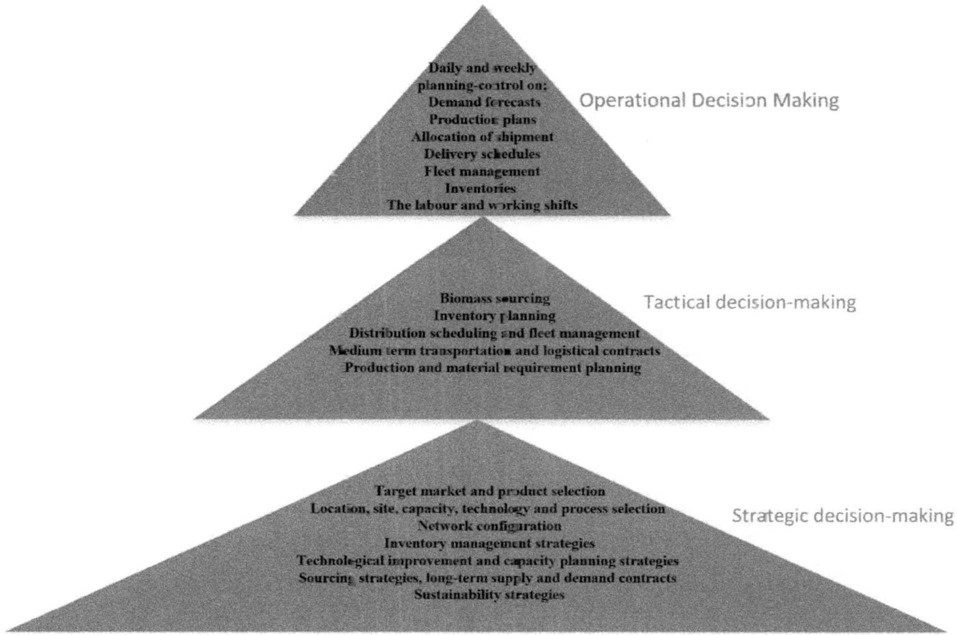

FIGURE 6.1 The hierarchy of decision-making process for the design, planning, and management of biomass-based production chains.

of biomass-based production chains. In the design phase, the biomass suppliers, distributors and service suppliers, customer service areas and associated transportation links (i.e., network decisions), should be selected and optimized, so as to minimize the material distribution and transportation costs from the preselected sites of conversion facilities and storage. Optimized supply chain networks ensure the reliable and efficient delivery of biomass sources and bio-products from supplier to the end user. For enhanced configuration decisions, the trade-off between transportation costs, which constitutes a significant part of total supply chain costs, and capital investment costs for facilities, should be considered, due to the fact that the decision-maker may encounter excess construction costs and land price (both are important elements of investment cost), for the facility sites that are close to the biomass and demand nodes. Construction costs and land price may change considerably according to different countries, even different regions in a country. To optimize the location and site for the biomass-based facility, specific governmental subsidies that change according to the region, the production volume, and the purpose and scale of the facility, should be considered. The same trade-off should also be considered while choosing between one or a few centralized storage facilities and multiple decentralized storage. The major advantage of multiple decentralized storage is lower transportation costs, due to the closeness to at least one supplier or demand node, against higher total investment cost. Another trade-off is between the capital investment cost and operational cost, which depends on the scale (or capacity) of the facility. To select the most appropriate location and site for a biomass-based conversion, storage or pretreatment facility, the distance of candidate locations to transportation links and infrastructure, biomass source sites, water bodies, and residential areas should be taken into account. Distance to residential areas is a significant factor to minimize the negative impacts of the facility on the nearby population (e.g., bad odors) and to minimize the transportation and distribution cost of bio-products. Air distance might be an important parameter particularly when construction of a pipeline system to transfer bio-fuel or heat is the case. To address the optimum capacity for a conversion plant, pretreatment facility or storage, economies of scale should be taken into account. Making strategic decisions in supply chain management requires providing the equilibrium among investment, operation, transportation, inventory, and procurement costs by the suitable optimization procedure that preferably covers the entire network, in order to improve the service levels and overall profitability, simultaneously. One of the most effective tools for visualizing the geographical allotment and spatial distribution of sites for significant biomass sources, suitable areas as candidate sites to locate facilities, demand nodes and densities that are considered to be parameters in the modeling process, is the geographic information system. For detailed information about geographic information systems, refer to Chapter 7 of this book.

Another critical strategic decision is, the selection among a number of options for technologies and processes, and required infrastructures for conversion of waste biomass and organic substrates into a wide range of bio-products, treatment, and storage of biomass sources and bio-products. Since this type of decision will not be changed in a short-term period due to high capital and set-up costs, technology and process decisions should be optimized at the beginning of planning the generation of biomass-based products. Different technologies may significantly vary in the levels of productivity, quality, and yields for biomass-based products, the appropriate biomass resources to be used, as well as the capital and operational costs. The key points to designate the most convenient technology and process for the facilities are, the amount, type, and spatial distribution of available biomass sources, end user applications and the specific characteristics and quantity of demand, cost of construction and maintenance of equipment, the level of environmental emissions and pollution from the production process, and local policies and restrictions at the facility sites.

Technology and process selection problems can be explored at two levels, the supply chain level and the facility or process level. The main issue is the selection of conversion pathways that specify the required technologies and processes at the supply chain level, whereas the subproblems of specifying equipment and catalysts, determining operational conditions and characterization of the specific process requirements corresponding to the demand type, are addressed in the process level.

The terms for purchasing and delivering biomass sources and bio-products agreed between supplier, producer, and end user are guaranteed by supply and demand contracts. These contracts might either involve standard regulations such as investment subsidies, tax reductions and exemptions, incentive schemes for bio-energy and blending restrictions for bio-fuels, or specific terms that regulate the material and financial flow between contractors. Supply and demand contracts ensure both the supply of material with adequate quality and quantity to fulfill the demand, and complete and timely payment under reliable and stable conditions. The contracts with suppliers (generally in agricultural sector), governmental units, and end users should provide the basis of biomass-based production, transportation, and consumption policies and a collaborative effort should be organized, to efficiently integrate a wide range of expert opinions from each team, on the ways of supporting the key functions of the supply chain and achieving the overall objective. Outsourcing, which means collaborating with outside companies that have an expertise in certain functions such as accounting, payroll processing, logistics, and inventory management, to handle the work typically performed by the firm itself, may be an option for many companies to cut down their costs. Transferring a part of the work to a company specialized in a field may provide increased efficiency, reduced labor cost, flexibility, and risk reduction. The sourcing strategy also has a significant influence on inventory and product management strategies, including management of on-farm and in-forest storage, and buffer stocks in conversion and preprocessing facilities, which should be determined, considering the type and specific characteristics of resources and products, as well as the products' and supply chain's life cycle.

In strategic planning, the possible economic, environmental, and social impacts of the supply chain and its activities should be addressed adequately. A variety of strategic decisions in biomass-based production chains should be made, so as to ensure the entire supply chain has a sustainable configuration and operations, to accomplish the main functions in the life cycle of the supply chain. For example, the technology, location, and site selection decisions for the conversion, treatment, and storage plants should be made by analyzing the potential social and environmental effects of the facility to mitigate undesired consequences for the local people and surrounding ecosystem. A detailed information on sustainability issues in biomass-based production chains is provided in Chapter 4.

Although strategic decisions are long-term and it is not easy to change them in the short-term, there is a possibility of future adjustments and improvements to accommodate the future fluctuations in demand and other market conditions. The main adjustments of this kind in the supply chain include, replacing the system in another location or site, capacity expansion or reduction, technological improvements and changes, and changing the terms of the supply and demand contract and even canceling it. Unstable conditions in the existing markets may force investors to find opportunities to expand to new or developing markets. This situation may create a need to produce modified versions of the existing products, or new products to meet the new market's demand, which results in the alteration of the existing technology or process pathways. In this case, the technological equipment or system may be exposed to modification, adaptation, improvement processes, or it even may be completely changed. Adaptation to new markets may also require changes in supply and demand contracts and sourcing strategies. Another reason for adjustments in strategic decisions is the budget limitation for investments. In cases where the total cost of investment exceeds the budget estimated in the planning and design phase, a number of strategies may be followed including, reducing material processing, treatment, or conversion capacity of one or more facilities, modifying inventory strategies so as to cut down the storage and material holding costs, reducing material distribution costs by introducing new transportation vehicles, modes, and strategies, improving product costs by renewing the raw material supply or demand contracts, or evaluating different supplier alternatives. Furthermore, the configuration and networking decisions may evolve over time due to the improved incentive schemes for renewable-based production, technology upgrading, and the wider acceptance of biomass-based products. These developments in the biomass-based production sector may stimulate capacity expansion decisions. To create a flexible decision-making procedure that allows for adjustments and modifications to strategic decisions, multi-period planning models should be adapted that divide the planning horizon into several strategic periods.

6.1.2 Tactical Decisions

Tactical decisions are medium-term decisions usually spanning between six months and one year. Tactical decisions are made within the limitations of the overarching strategic supply chain decisions, in other words, they are planned towards accomplishing and executing the strategic level decisions. Tactical level decisions aim at creating medium-term schedules, and plans with the aim of better utilizing the configuration established in the strategic level decision-making, and at the same time identifying medium-term policies related to production, material planning, distribution, and storage. Furthermore, tactical level decisions help in ensuring efficiency in production, distribution, and other operations that add value to the product and to maintain a specified level of service and product quality in a cost-effective way.

Capturing tactical level decisions and strategic level decisions simultaneously, in configuration and network design problems, lead to improved solutions in terms of optimal network infrastructure that enhance the usage and allocation of production, storage, and distribution resources to meet the end users' demand with a specified service level and cost. Common tactical decisions in biomass-based production chains include:

- biomass sourcing and procurement of auxiliary material,
- inventory planning and warehouse management,
- distribution scheduling and fleet management,
- medium-term transportation and logistical contracts, including outsourcing and third-party options, planning the transportation modes and vehicles to be used,
- production and material requirement planning (MRP),
- selection of cultivation, harvesting, and collection methods for biomass sources,

Decisions related to biomass sourcing and procurement of auxiliary material are important in biomass-based production chains to minimize the geographical distances between the suppliers and conversion, pretreatment, or storage facilities and increase availability to the biomass sources, which are the key factors in minimizing transportation and purchasing costs. Decision-making methodology, to make biomass sourcing decisions, should involve the identification of the most accessible biomass types and related biomass quantities in a given region, to select the optimal biomass sources and associated supply quantities, as well as the biomass suppliers. Once again,

Geographic Information Systems (GIS) can be used to visualize the spatial distribution of sites for significant biomass sources over a region, besides the density of the biomass source at related site.

Production and MRP decisions are among other types of significant medium-term decisions. MRP decisions ensure that the lowest possible amount of necessary materials are available for production processes, and the quantity of final products is sufficient to meet the market demand, while maintaining the optimized material and product levels in stock to avoid shortages. In a biomass-based production chain, the efficiency of the MRP system impacts the performance of conversion, pretreatment and storage functions, distribution schedules, and material purchasing. MRP constitutes a basis for a production planning system, which makes plans for future production over a specific planning horizon, to create a wide range of products for different end users, determining the required facilities and allocating labor and other resources to activities, so as to use the maximum level of productive capacity of available resources. The main purpose of production planning is to decide the optimum product mix consisting of a variety of product types, and setting production/delivery timing and orders, considering the urgency of different customer needs. While organizing the production, it is important to match the existing resources to the required level of production with the highest efficiency. Accuracy and reliability of a production plan depend on the data gathered in collaboration with purchasing, marketing, sales, and inventory management modules, such as material and resource availability, customer orders, sales forecasts, and inventory levels, corresponding to the scheduled products. An efficient production plan optimizes both production system-related performance measures (e.g., cycle time and lead time) and customer satisfaction-related measures (e.g., delivery time).

Inventory planning and warehouse management decisions can be considered as key tactical level decisions to facilitate the effective and timely stocking of raw materials or finished products. The main objective of the inventory planning and warehouse management activities is, to develop and employ strategies to optimize the inventory levels, to maximize the service level and level of demand fulfillment through preventing stock outs and backlogs, and to minimize the cost of holding stock. To provide the equilibrium among inventory costs and service levels in terms of stock outs and backlogs, and to make use of the locked-up working capital in the warehouse, efficient inventory management techniques should be used. The availability of raw material for conversion and pretreatment processes as well as finished products is critical to accommodate the unexpected variations in market demand of products, which is one of the most important factors to maintain the flexibility and responsiveness of a supply chain. Besides the availability of raw material, an overall supply chain planning strategy involving the interaction between biomass sourcing, production planning, and inventory management phases, is the key to obtain an optimized level of stored raw material and finished products. In addition, if we consider a multiperiod structure in the tactical decision-making and modeling process, the optimized production level in each time period depends on the inventory level in the previous time period.

Decisions related to transportation and distribution of biomass sources, bio-products and other material are typically made to regulate the flow of goods from one location to another to satisfy customers' demand in a timely and cost-effective manner. This type of decision deals with scheduling of distribution activities, fleet management, determination of delivery amounts to each demand location, selection of transportation mode and vehicles, specifying delivery routes, and medium-term contracts with transportation and logistics providers. In some cases, outsourcing may be a reasonable way to reduce certain costs associated with logistics systems, such as, costs caused by governmental mandates and excessive taxes, energy consumption, or labor. Outsourcing allows a company to transfer part of its activities to a third-party firm. Activities that a company contracts out may be associated with the core functions of the production system (e.g., conversion or pretreatment processes), noncore functions (e.g., transportation, facility management), financial management (e.g., accounting, debt management), or supportive functions (e.g., maintenance of equipment, customer services). To make enhanced decisions related to transportation and distribution systems, data about demand and sales forecast, material supply, inventory as well as lead time and due date restrictions should be collected in collaboration with marketing and sales, production planning, and inventory management modules.

6.1.3 Operational Decisions

Operational decisions are short-term decisions that are made generally weekly, daily, or hourly, focusing mainly on the details of operations, day-to-day resource allocation, details of inventory control, and delivery routing, to ensure the efficiency of operations and an optimized flow of products along the biomass-based production chains. Operational decisions can be revised and adjusted frequently, due to the dynamic structure of internal and external conditions of supply chains and associated activities.

Operational decisions are made to execute the short-term processes with the aim of achieving the long- and medium-term goals that the strategic and tactical level decisions have adopted. Given the constraints established by the configuration and medium-term planning policies, the goal during the operational level is to guarantee the continuous operation of the facilities in the supply chain and to maximize benefit from short-term activities. These activities include:

- daily and weekly forecasting to estimate the end users' demand,
- setting the due date when demand is to be met and reviewing the variations between lead time and due date,

- daily and weekly production planning and scheduling,
- allocation of a shipment of biomass sources or products to a particular transportation mode and vehicle,
- daily and weekly delivery schedules and routes, fleet management,
- monitoring shortage of supply and backlogs, planning additional activities in daily and weekly basis to meet the shortages,
- daily and weekly inventory control and replenishment,
- organizing the labor and working shifts.

The operational level plans at the biomass acquisition stage involve, allocation of the operators and other resources (such as fuel) to harvesting machines, assignment of the harvesting operation areas and setting due dates to harvest each area, determining the way of conveying the harvested biomass to storage, and preprocessing units with the corresponding routes. The detailed inventory planning policies determine the frequency of inventory replenishment with the supply of biomass from the upstream of the supply chain and frequency of delivery from the storage to the downstream of the chain. Besides the frequencies, the quantity of the biomass and bio-products to be delivered to and from storage can be determined on a daily and weekly basis. These decisions may be more challenging if multiple types of biomass sources are utilized. In addition, the seasonal variations in biomass acquisition and supply often require more comprehensive methodologies with uncertainty consideration for operational level decision-making.

At daily and weekly production planning and scheduling, the main operational decisions comprise scheduling of multiple tasks for production operations, allocation of operators to the operations, and conversion and preprocessing equipment, the coordination of material requirement and flow, with the aim of ensuring the timely production and delivery of demanded products in prespecified quantity and quality. Although MRP decisions are considered among tactical level decisions, in operational level, daily and weekly reviews of the medium-term MRP decisions should be conducted so as to monitor the unexpected fluctuations in demand, material supply, or operational conditions, and to revise the decisions to accommodate these fluctuations and minimize their negative impact on the production and distribution systems. The values of operational parameters for conversion processes, such as pressure, temperature, and flow rate, significantly depend on the feedstock characteristics (e.g., moisture content, total solid content, calorific value). To capture the operational level problems in production processes, the physical and thermodynamic properties of biomass should be exploited by different methods (e.g., dynamic modeling, simulation, advanced control algorithms).

Operational decisions in the biomass-based production chains impact the timeliness, efficiency and cost-effectiveness in material flow, conversion/pretreatment operations, and storage. The tools and methodologies developed, to be used in making the above-mentioned detailed short-term decisions in generic supply chains, can also be adapted to biomass-based production chains, such as enterprise resource planning (ERP) and MRP tools, demand forecasting models based on time series, material allocation procedures, service level measures, and information flow diagrams.

Another decision that is frequently revised, due to the variations in the lead time as a result of production and operational changes, is the due date that demand is to be met. In this scope, tardiness and earliness measures are used in scheduling. Tardiness defines executing operations with a delay, whereas earliness is a measure of completing operations before the due date. Tardiness may be caused as a result of problems of availability of raw materials or equipment, or the dependency of the jobs or processes on each other. In biomass-based production systems, both tardiness and earliness may be undesirable in terms of inventory costs and customer satisfaction. Earliness involves additional expenses for the storage of bio-products and may be a reason of degradation in certain types of biomass. Tardiness may result in stock outs and backlogs and therefore, delay in the fulfillment of customers' demand, which leads to loss of commercial standing.

Demand forecasting aims to estimate future demand of a product through historical data and experiences. Appropriate demand forecasting techniques help in gathering information on current and potential markets of a business, to make strong decisions about products and target consumer groups, so as to determine pricing and capacity expansion strategies. Also, demand forecasting has a significant influence on inventory management, allowing optimized inventory levels and turnover rates, and minimized inventory holding costs. Efficient demand forecasting and management enhances allocating a sufficient amount of resources to production activities, that is, increasing required labor and other resources to keep production processes running smoothly and accommodate the increase in demand during peak periods. In addition, the future cash flow can be estimated more reliably and businesses can give more accurate budgeting decisions to cover the operational and other expenses.

6.2 DECISION-MAKING IN DESIGN AND MANAGEMENT OF BIOMASS-BASED PRODUCTION CHAINS

A wide range of objective functions and optimization criteria enables expressing different aspects, attributes, and business goals of biomass-based production systems in mathematical terms, for utilization in decision-making procedures, which comprise operations research and optimization techniques. Criteria refer to a set of requirements or attributes (properties), used to evaluate the performance of processes or alternatives, to establish a reliable judgment regarding to the objective achievement, whereas the objective can be defined as the purpose, target, goal, or standard, in other words, the desired state of the system, related to or derived from a set of attributes that should be accomplished by employing predetermined business strategies. Depending on the characteristics of the problem, the same measures can be used to represent the objectives and criteria. For example, minimization of capital investment costs can be used as the objective function in a supply chain design problem, however, the same measure can also be taken into account as one of the criteria to evaluate more than one configuration alternative in another problem. The multiple attribute decision-making (MADM) and multiple objective decision-making (MODM) methodologies integrate a set of criteria with different priorities and conflicting objectives to obtain a sustainable supply chain network configuration and operational efficiency. This section covers; (1) the main economic, environmental, social, and technical objectives and criteria, and (2) MADM and MODM methodologies, that can be used to handle design and management problems comprising multiple criteria and goals to represent various economic, environmental, social, and technical aspects of biomass-based production chains.

6.2.1 The Main Objectives and Criteria

This section explains the objective function structures and criteria that have a wide application area in scientific studies and practical cases to develop and apply analytic methodologies for modeling, optimization, design, and management of biomass-based production systems and supply chains. Some additional objective functions, constraints, and criteria, specific to the problem to be handled, may require being captured depending on the case, however, this section focuses on the most widely used structures. Information on some other criteria (i.e., sustainability criteria) can be found in Chapter 4, sustainability issues in biomass-based production chains.

6.2.1.1 Economic Objectives and Criteria

Economic objectives are the economic targets set by the decision-maker, either to design and optimize an economically sustainable and profitable system, or to maintain an optimized level of competitiveness and flexibility by providing operational efficiency and regular cash flows in expected time and amount. Economic criteria can be utilized to assess the economic performance of different alternatives regarding supply chain configuration or an element of the chain (e.g., a system, a plant, or a process). Economic performance of the biomass-based production chains can be measured by several metrics, which can be used as objective functions or criteria, depending on the type and purpose of the problem (i.e., optimization problem or evaluation/assessment problem), can be categorized as costs, revenues, profitability, net present value (NPV), return on investment (ROI), internal rate of return (IRR), payback period, and benefit/cost ratio. These metrics are generally determined considering the time value of money over the lifetime of the production system or supply chain elements (typically 15–20 years for biomass-based production chains).

Overall cost or components of the overall cost are commonly used as minimization objectives or criteria in the modeling and design of biomass-based production chains. The components that form the overall cost in a supply chain can differentiate between various optimization problems and have a strong relationship with the planning decisions and prominent supply chain activities. The basic elements of the overall cost can be investigated in two classes, strategic level costs, associated with the design and establishment of the supply chain elements and network configuration, and tactical/operational level costs associated with supply chain activities. Strategic level costs include the capital investment cost of facilities (i.e., conversion, pretreatment, and storage facilities), which comprise the equipment and building costs, land cost, construction, and associated project management costs. The tactical/operational level costs involve annual operational and maintenance costs (including cost of labor), in conversion and pretreatment plants, biomass and auxiliary material purchasing costs, biomass and bio-product transportation costs, biomass and bio-product inventory costs, annual costs for cultivation, harvesting, and collection of crop-type biomass. To calculate the overall cost of the supply chain accurately, the capital investment cost should be discounted using a proper interest rate over the lifetime of the system, to the same time period as the other cost elements (e.g., annual or monthly). Fig. 6.2 depicts the main elements forming the overall cost of biomass-based production systems. In addition to the cost elements depicted in Fig. 6.2, some different problem specific costs may also be considered in varying cases.

Another option for expressing the economic objective or criteria is the total supply chain profit that has a wide range of application areas in the modeling and design of biomass-based production chains. The supply chain's total profit is determined, based on total revenue minus total cost. The core driver of the studies related to biomass-based production system design and management, with the objective of the maximization of total supply chain profit, is to obtain a supply chain configuration and a set of activities and processes so as to minimize the overall costs and simultaneously maximize the total revenues. In biomass-based production chains revenues are usually generated by the sale of a wide range of bio-products, which are the main products of the system (e.g., bio-fuels, bio-energy, bio-chemicals). However, for a more effective decision-making process, one should also take into account the government incentives and subsidies and, additional income from selling the by-products such as bio-fertilizer, bio-char, and other residual products that can be used in chemical industry.

Overall cost and total profit are the most widely used economic indicators as objectives in optimization problems and criteria in ranking/selection of investment alternatives, and performance evaluation studies. However, some

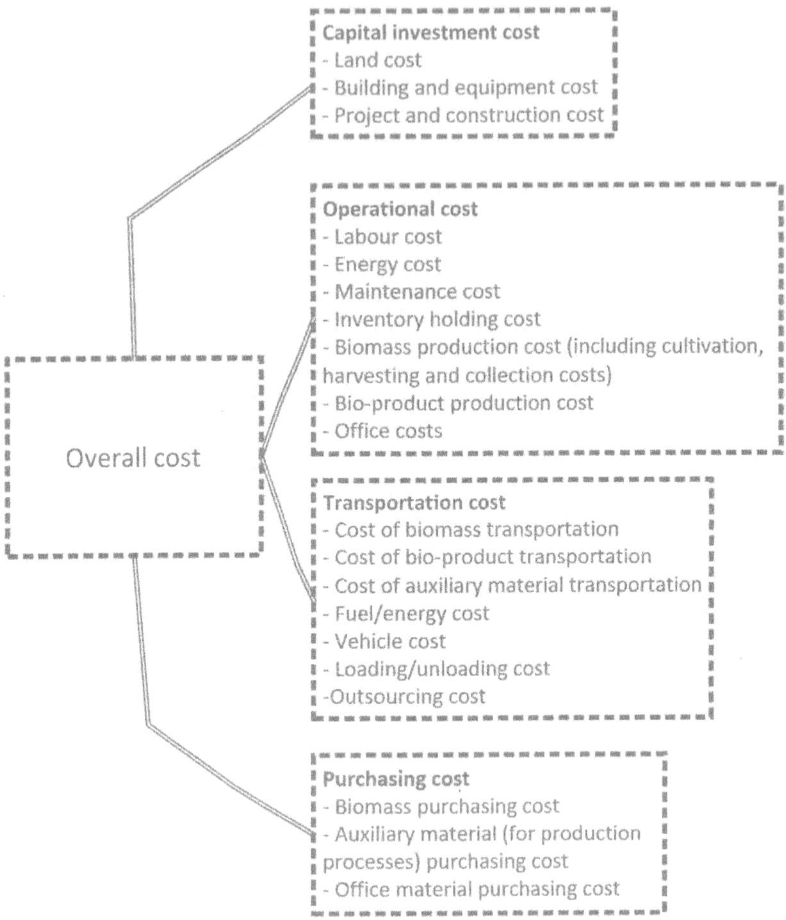

FIGURE 6.2 The main elements of the overall cost of biomass-based production systems.

other economic performance measures can also be used to characterize the economic objective function and criteria, such as NPV, ROI, IRR, and the payback period. Among them, ROI and NPV are indicators of profitability of a system or supply chain, having the advantage of ease of use, adaptability, and practicality. NPV can be defined as "the sum of benefits gained, minus the sum of costs endured". To calculate NPV, all costs (including initial investment cost) and benefits encountered in each time period, during the lifetime of the investment, should be discounted to the present time considering the periodic interest rate. In NPV calculations, the time value of money is taken into account, which dictates the impact of time on the value of cash inflows (i.e., benefits or revenues) and outflows (i.e., costs or expenses). The time period in biomass-based production chain design and modeling studies is typically one year, but could also be quarter-years, half-years or months.

NPV can be used as either an objective function or an evaluation criterion, to determine whether an investment will result in a net profit (a positive NPV) or a loss (a negative NPV). The NPV of an investment is desired to be maximized in optimization studies, and in investment appraisal, the option with the highest NPV is chosen among a number of investment alternatives. ROI identifies the percentage of money gained or lost during the lifetime of an investment to the total amount of capital investment cost. The project lifetime or planning horizon and interest rate are the factors that influence ROI as economic criteria. ROI is typically used either as a metric in investment decision-making, identifying the rate of return on money spent to perform an operation, or to buy an economic entity, or to compare different investment options. The option with the largest ROI should be chosen.

The IRR is another economic objective and criterion that can be used to compare the profitability of different investment alternatives. IRR refers to the discount rate at which the NPV of all future cash flows is equal to the initial capital cost of an investment, in other words, IRR is the discount rate that makes the NPV of negative cash flows (i.e., costs) equal to the NPV of positive cash flows (i.e., benefits). In the case of selecting between several investment alternatives, the higher an investment's IRR, the more desirable it is to undertake the investment, and in the case of evaluation of a unique investment, the IRR should be higher than the minimum acceptable rate of return, to accept the investment.

To measure how long an investment takes to pay for itself, the payback period is utilized, which refers to the period of time to cover the expenses of an investment or to reach the point that the company has no net profit or loss (i.e., the break-even point). The payback period is generally expressed in years. The time value of money, that changes due to variations in earning power of money or the inflation rate over time, is not taken into account in a payback period calculation, however, it is a popular economic criterion because it is easy to understand and apply for most individuals. The investments with shorter payback periods are preferable to the ones with longer payback periods. It is commonly used to compare different investment alternatives, such as conversion technologies, in biomass-based production investments, however, when it is used in isolation to compare an investment versus a "doing nothing" option it is not a reliable decision-making criterion. Due to this important drawback, it is generally not recommended to use this criterion as standalone for investment decision-making. Instead, a discounted payback period, by which the time value of money can be rectified by discounting future cash flows, can be used. The discounted payback period refers to the period of time it takes to break even from the inception of investment, considering discounted cash flows. Another major drawback of the payback period is that it does not take into account the cash flows after the payback period of the investment, however, most biomass-based investments have a long service life and cash flows occur beyond the payback period. Hence, the payback period can be seen as an indicator of the short-term profitability, and a valuable project may be ignored if the payback period is the only consideration that it should be used, along with other economic criteria such as NPV or IRR. The payback period is mainly used for financial evaluation, however, it is occasionally extended to other areas. As an example, an energy payback period gives a time period to cover the energy expended since undertaking the initial expenditure by the energy savings.

When using as objective function in a mathematical model, NPV, ROI, and IRR should be maximized, whereas the payback period should be minimized. These four metrics can be used, for example, to compare the profitability of each project of an investment in a new conversion plant with an extension of an existing plant. If all projects require the same amount of capital expenditure, the project with the highest NPV, IRR, and ROI would be selected as the best and be undertaken first. However, it is important to consider the possible budget constraints that may restrict the selection process or the projects that are interdependent and which should be executed consecutively.

6.2.1.2 Environmental Objectives and Criteria

Environmental objectives are defined, to design and plan the biomass-based system and related operations, so as to minimize the harmful impacts of the system on nature and the surrounding ecosystem, whereas environmental criteria can be used to evaluate the environmental performance of different systems, processes, or technologies according to different impact categories. The main difference between economic and environmental objectives and criteria is that the total quantity is optimized for most economic objectives (e.g., minimization of total cost, maximization of total profit), whereas the "environmental impact per functional unit" is more critical for environmental objectives and criteria (Yue, Kim, and You, 2013). For example, minimizing the carbon emissions resulting from the production of 1 m^3 biomethane may be taken into consideration, rather than minimizing the total annual carbon emissions resulting from all conversion operations throughout the supply chain. However, using fractional terms to formulate functional unit-based environmental objective functions may result in nonlinear equations, which require further linearization methods, advanced solution algorithms, or nonlinear programming solvers. Life cycle environmental impacts to air, soil and water, balance of greenhouse gases (GHGs), water-use, and land-use are the major environmental considerations that are represented in objective functions and criteria set in the quantitative studies, to model and optimize biomass-based production chains. To quantify the environmental impacts of the biomass-based production systems in different categories, the "footprint" concept is widely utilized. The ecological footprint can be described as, a quantitative measure of human demand on natural resources required for the production of the main human needs, including food and feed products, energy, timber and other forest products, shelter and urban infrastructure facilities, and for disposal of its liquid, solid, and gaseous waste. The latter demand is highly correlated with the waste absorption capacity of nature. The footprints can be measured in units of area, however, defining some footprints may impose difficulties, especially for the processes that are not based on area, such as chemical processes (De Benedetto and Klemeš, 2009). Also conducting footprint measurements, to provide a basis for formulation and quantification of environmental objectives and criteria, may be challenging due to the uncertainty of environmental data and limited data availability. Carbon footprint, water and water pollution footprint, energy footprint, and agricultural land footprint are the major environmental footprints that are used to quantify and integrate the environmental impacts in procedures for design, planning, and management of biomass-based production systems.

Carbon footprint is one of the most important environmental performance metrics for biomass-based production chains, which can be defined in terms of the amount of carbon dioxide, or its equivalent of other GHGs, resulting from the activities of companies, individuals, production, logistics, and consumption operations in the chain. It may be quite challenging to calculate the total carbon emissions or the equivalent of other GHG emissions in a production chain, due to the lack of data about complex interactions between processes and facilities in the chain, which may comprise natural processes that both store and release carbon dioxide. For this reason, Wright et al. (2011), defined the carbon footprint as "a measure of the total amount of carbon dioxide and methane emissions of a defined population, system, or activity, considering all relevant sources, sinks, and storage within the spatial and temporal boundary of the population, system, or activity of interest". Limiting the type of emissions as carbon dioxide and methane leads to a relatively straightforward data collection. Carbon footprint can be quantified as a carbon dioxide equivalent using a global warming potential (GWP) indicator, which defines the amount of GHGs that has a contribution to the global warming (or climate change) in a specific time period, typically taken as 100 years (GWP100). Carbon footprint can also be measured in unit area in that it represents the land area required for the sequestration of fossil-fuel-based carbon dioxide emissions from the atmosphere through afforestation (De Benedetto and Klemeš, 2009). Apart from the carbon footprint, GHG footprint, methane footprint (Wiedmann and Barrett, 2011), climate footprint (Wiedmann and Minx, 2009), and GWP footprint (Meisterling et al., 2009) are among the other terms related to atmospheric emissions that have been suggested and/or used in recent years by a number of researchers.

Water footprint can be used to define basically two environmental metrics in the biomass-based production chains; (1) the total volume of direct and indirect freshwater used in all types of operations performed in the chain and (2) the water pollution footprint that can be defined as the amount of harmful substances emitted into water as a result of the supply chain activities. The first metric is used for quantifying the total water consumption for a particular operation, for a product, for a specified group of consumers, or stakeholders in the supply chain. Water footprint can be divided into three categories according to the source of water consumed (incorporated into the product or evaporated) in the operation, blue, green, and gray water footprints. Blue water footprint refers to the consumption and evaporation of surface and groundwater resources. Irrigated agriculture, industry, and domestic water-use can each have a blue water footprint. Green water footprint represents the consumption of rainwater, especially that stored in the deepest part of the soil and evaporated, transpired, or incorporated by plants. It is especially suitable for agricultural, horticultural, and forestry products. Gray water footprint is the volume of water needed to assimilate pollutants to upgrade the water, according to specific water quality standards, by directly discharging the pollutants from contaminated water through a pipe or indirectly through leaching from the soil. Water footprint analyses provide a deep understanding on the water usage and water management in a system, however, the related environmental impacts of different types of water footprint should also be revealed besides the quantity of water used.

A number of definitions have been offered for energy footprint, which can be divided into two categories, as land-based and energy demand-based definitions. Energy footprint can be defined as the metric that indicates the demand for nonrenewable energy resources, according to Schindler (2010). Energy footprint is defined by The Global Footprint Network (GFN) (2009), as the sum of all areas used to provide non-food and non-feed energy, such as land used for hydropower, cultivated land for energy and fuel crops, and land for forest wood fuel. GFN (2009) also described local and global units for energy footprint; it can be measured locally, considering the surface area of a region's average biologically productive land and sea over a given year, and globally, considering the surface area of the earth's average biologically productive land and sea over a given year. Another land-based definition of energy footprint is that it refers to the total area of land that would be needed to absorb the carbon dioxide emissions from energy consumption (Palmer, 1998). Energy footprint can be investigated, focusing on either renewable energy (Chen and Lin, 2008; Santhanam, 2011; Brown, 2009), or nonrenewable and fossil sources-based energy (Stoeglehner and Narodoslawsky, 2009; Stoeglehner et al., 2005).

To quantify the environmental impacts of biomass-based production systems on air, water, and soil, emission footprint has been introduced. Emission footprint is calculated per unit of area. Emissions to the air typically indicates the quantity of SO_2, CO, CO_2, and particle emissions. Chemical Oxygen Demand, which is a measure of the oxygen consumption capacity of water during the decomposition of organic matter and the oxidation of inorganic chemicals, can be used to quantify the water footprint of a system as well as nitrogen and phosphorus emissions to the water. The total amount of waste disposed of improperly through spillage in the soil can be used to quantify the emissions into the soil. The maximum amount of emissions into the air, water, and soil that would not have an impact on the qualities of the local compartments, can be calculated and compared with natural flows, based on the naturally existing qualities of these compartments and the rate of replenishment per unit of area. The rate of replenishment can be determined, based on the natural exchange of substances between forests and air per unit of area, the seepage rate calculated, based on local precipitation rate, and the decomposition of biomass into humus calculated, based on the conversion rate of biomass into compost, for air, water and soil, respectively (Cucek et al., 2012).

Biomass (other than residue and waste-based biomass) has a significant land footprint, which means that land-use and land-change issues, related to biomass-based production, need particular attention and evaluation. In the concept of biomass-based production, land footprint can be defined as, the amount of land that is needed to produce a bio-product or biomass source or is used by a company/stakeholder/individual in the production chain. The land footprint involves different kinds of sub-footprint according to the type and purpose of the land used in different stages of the supply chain. Among them, agricultural land footprint (Kissinger and Gottllieb, 2010) and forest footprint refers to the land used in the biomass production stage, which represents, respectively, the land used to cultivate biomass sources and the area required to produce the forest-based biomass sources. Other types of land footprint are the built-up land footprint (Chambers et al., 2004), which are lands occupied by infrastructures used by humans and the grazing land footprint that is the land used for livestock (GFN, 2010). Land footprint of a biomass-based system is highly correlated to land-change, which is defined as the change of the land from one type to another. Selection of the size, capacity, and scale of the biomass-based production/conversion units, as well as the agricultural area to grow biomass crops should be handled by proper optimization methodologies to avoid the use of excess land.

6.2.1.3 *Social Objectives and Criteria*

In general, modeling and optimization studies on design and management of biomass-based production chains focus on two social dimensions in objective function structures and criteria set, of which, measurement, formulation, and quantification are easier in comparison with other social indicators, job creation, and contribution to the local employment pattern and food security.

Job creation and contribution to the local employment pattern can be represented by two metrics, direct employment and local employment.

Stamford and Azapagic (2011) developed the formulation given below, to measure the total number of jobs on a life cycle basis directly in a gas to power conversion system, based on the number of jobs created along the supply chain, the duration of employment, and the total amount of power generated over the lifetime of the conversion plant:

$$DE = \frac{\sum_{i=1}^{I} DE_i \times t_i}{P_{tot}}$$

where, DE_i represents the number of jobs created in stage i of the life cycle, t_i represents the duration of employment in stage i, and P_{tot} denotes the total amount of power generated over the lifetime of the conversion plant. The results belonging to each stage is calculated to find the life cycle-based total direct employment generated, along the supply chain (person-years/TWh). This approach can also be used to quantify the direct employment generation impact of the biomass to bio-product conversion systems, by modifying the denominator of the formulation to "the total amount of bio-product (in different units such as tons, liters, or m^3), generated over the lifetime of the conversion plant".

Contributions of the biomass-based production chain to the local employment pattern can be represented by employment of the workforce from the area, where the supply chain or conversion plant is located, including the neighboring communities. Local employment is defined by Stamford and Azapagic (2011), as the percentage of new jobs created that could be occupied by employees from local communities. Contributions to the local employment pattern can be quantified using the following equation:

$$P_{LE} = \frac{LE}{TE} \times 100$$

where, P_{LE} denotes the proportion of employees that could be hired from the local community (%), LE is the number of employees that could be hired from the local community (persons-years/TWh), and TE is the total number of employees needed (persons-years/ TWh).

Food security is an important social concern, especially in biomass to energy and fuel conversion systems, and should be handled in optimization and modeling of biomass-based production chains. Food security is highly related to land-use and land-change issues, since the change of agricultural lands for growing food crops into biomass production fields for energy, fuel, and other purposes, may result in a competition between the food and energy/fuel sectors. Hence, the location and size of the agricultural lands for biomass crop production purposes should be optimized to avoid such a competition. Also, the use of food crops (e.g., corn, sugarcane), grain, sugar, and oil crops as feedstock in the production of bio-fuels, such as bio-diesel and bioethanol, may threaten the safety of food supplies and sustainable food production. Hence, the size of the agricultural area changed from food production purposes to biomass for energy/fuel production purposes should be optimized, and excess use of these lands should be avoided. Also, the amount of food crop used for energy/fuel production should be minimized, to design a supply chain that does not threaten the security and sustainability of food resources. The food to energy footprint, which can be measured by the mass flow-rate of food-intended crops converted to energy, can be used to assess the possible competition between the food and energy sectors (Cucek et al., 2012).

Social return on investment (SROI) can also be used as a social objective or criterion in the design and management of biomass-based production chains, which provides a way of incorporating the environmental and social values, that are not currently reflected in conventional financial analysis, into decision-making procedures. SROI presents a systematic way to quantify the social and environmental outcomes of an investment related to resources invested. SROI uses monetary values to represent the social, environmental, and economic outcomes of an investment and shows the ratio of benefits to costs (e.g., a ratio of 2:1 indicates that an investment of £1 delivers £2 of social value). Among other social issues, social acceptability and labor conditions can be included in the decision-making process using linguistic variables, or can be quantified by calculating the number or percentage of supporting people in the local community (i.e., the community in the area where the supply chain is located including the neighboring communities), and the number of worker injuries, respectively.

6.2.1.4 Technical Criteria

The technical and technological dimension is generally represented in the constraint set rather than an objective function in the mathematical programming approaches. There are three major technical considerations that limit the productivity of the biomass-based production system, and impact the optimization and modeling decisions in design and management of biomass-based production chains, the scale and capacity of the production system, the efficiency of the conversion/production technology/process, and the economic lifetime of the system.

The biomass-based production system can be small-, medium-, or large-scale. The medium- and large-scale systems are generally used for industrial purposes, whereas the small-scale systems can also be utilized in residential applications. The larger the scale of a system, the higher the capacity of the production and conversion units. The capacity of the production system can be input- or output-oriented, which means that it can be measured either by the maximum amount of biomass feedstock that can be processed by the system in unit time (e.g., 100 tons of dry biomass per day), or the maximum amount of final product that can be obtained by the system in unit time (e.g.,

15 l of bio-diesel per day). The capacity of the system imposes a considerable limitation on the yields and hence on the revenues. Also, it significantly impacts the investment costs, the large-scale systems with large processing and production capacity equipment result in higher initial investment costs, but lower investment and operational cost per unit of output because of the economies of scale, which refers to the cost advantages that enterprises obtain, due to the reduced costs per unit that arise from increasing scale and increased total output of a product. Hence, the production yields and accordingly revenues generated in a system, the capital investment cost, and operational and investment cost per unit of output should be considered simultaneously and compromised in design and management of the biomass-based systems. The initial investment cost may be restricted by a budget constraint.

The efficiency of the conversion/production technology/process poses a limitation on the system productivity and specifies the quantity of the final product per unit of input. For example, the efficiency of an anaerobic digestion system for biomass to biomethane production is determined by the technical characteristics of the digester, as well as the properties of the biomass material. Or, the efficiency of biomethane to energy conversion is specified by the technical properties of the cogeneration equipment and characteristics of the input biomethane, for example, 85% total efficiency, of which 77% is electrical efficiency, means that up to 85% of the biomethane fed into the cogeneration system can be converted into useful energy (electrical and thermal), of which 77% is electrical energy.

The lifetime of the biomass-based production system, equipment, or technology is an important factor in economic analyses of the investment that the costs and revenues, which are used to analyze the economic impacts of the system and included in the model as the components of economic objective function of an investment alternative, should be examined over the lifetime of the system. In this scope, the present worth of future costs and revenues should be calculated by a discounting process, using proper interest rates to compare different investment options related to system, equipment, or technology. When comparing mutually exclusive alternatives regarding an investment, there may be two situations that affect the analysis approach; (1) the analysis period (or required service period of the alternative) equals the lifetime of the alternative, and (2) the analysis period differs from the lifetime of the alternative. In the second situation, if the lifetime of the alternative is shorter than the analysis period, replacement investments would be required by renewing the equipment at the end of the lifetime. If the lifetime of the alternative is longer than the analysis period, then the equipment would be sold for its salvage value. Salvage value is the amount of money, for which the equipment could be sold, after its service to the project has been rendered. Alternatively, salvage value is the monetary measure of the remaining usefulness of the equipment.

6.2.2 Multiple Attribute and Multiple Criteria Decision-Making Methods in Biomass-Based Production Chains

The main difference between MADM and MODM problems is that, MADM involves the selection of the best alternative by considering the trade-offs within a set of interacting design attributes, whereas in the MODM approach, the decision alternatives are not given, and instead, MODM provides a mathematical framework for making a set of decisions.

6.2.2.1 Multiple Attribute Decision-Making

Roy (1996), defined the MADM as "a decision aid and a mathematical tool allowing the comparison of different alternatives or scenarios according to many criteria, often conflicting, in order to guide the decision-makers towards a judicious choice". Different attributes (or properties) of investment or project alternatives are used as classification criteria by MADM methodologies to accomplish mainly three goals; (1) to select the best or the most preferred alternative, (2) to rank the alternatives in descending order of preference, and (3) to sort out the alternatives that seem good.

In biomass-based production chains, the following elements can be assessed and ranked using MCDM methodologies:

- different biomass-based conversion processes to produce a particular bio-product, for example, gasification, combustion to generate power,
- different process/technology/equipment alternatives pretreatment to treat and/or upgrade the biomass feedstock or bio-product, for example, chipping, pelletization for biomass, biogas upgrade processes, bioethanol blending equipment,
- candidate locations and sites for conversion plants, storage, and pretreatment facilities,
- multiple equipment alternatives for a conversion process, for example, updraft, downdraft, and crossdraft gasifiers for gasification process or different Combined Heat and Power (CHP) alternatives with different efficiencies to generate heat,
- multiple feedstock options to produce a particular bio-product, for example, energy crops, municipal solid waste, and animal waste to produce biogas.

MADM methods provide a useful framework especially when there are many decision-makers involved in the decision-making process. The attributes of alternatives that form the criteria set may have different relative importance according to the preferences of the decision-makers. In addition, a criterion may be defined by linguistic values and/or quantitative values (e.g., the cost of a conversion system alternative is quantitative, whereas the social acceptability of the system can be defined by either linguistic values or quantitative values). In general, economic, environmental, and technical criteria are represented by quantitative values, whereas social criteria are more suitable for linguistic definition.

The analytical hierarchy process (AHP) (Saaty, 1980), is a well-known and commonly used MADM method that aids the decision-maker to set the priorities and determine the best alternative, through a series of pairwise comparisons using a set of evaluation criteria, and a set of alternatives. Both subjective and objective aspects of a decision can be captured and compared by AHP. Subjective aspects can be satisfaction feelings or preferences corresponding to an alternative, whereas objective aspects refer to the input that can be obtained from actual measurement (e.g., weight, price). In the first step of the procedure, a weight is generated for each criterion considering the decision-maker's pairwise comparisons regarding to the criteria. The higher the weight, the more important the corresponding criterion, in comparison with others. The next step evaluates each alternative option, with respect to each criterion, considering the decision-maker's pairwise comparisons to assign a score to each option. The options with better performance according to the considered criterion get a higher score. Finally, the last step involves combining the option scores and criteria weights, to assign an overall score to each option (i.e., the weighted sum of the scores corresponding to all criteria), that are used to rank the alternatives. Several commercially available decision-supporting software implement AHP, such as The Super Decisions, MakeItRational, and the Priority Estimation Tool. The analytic network process (ANP) is proposed by Saaty (1996), as a more general form of the AHP. AHP and ANP uses a similar implementation procedure, using a set of pairwise comparisons, to determine the weights of the criteria and to rank the alternatives in the decision problem, however, there is a major difference in that AHP models the problem, based on a hierarchy with a goal, attributes, and alternatives, while the ANP uses a network to structure the problem. Hence, ANP can be used to structure the decision problems with interdependent criteria, whereas each element in the AHP hierarchy is considered to be independent of the others.

VIKOR (VIseKriterijumsa Optimizacija I Kompromisno Resenje) was firstly introduced by Opricovic (1998), as a MADM method to solve the ranking, problematic in discrete multiple criteria problems, with noncommensurable and conflicting criteria (Opricovic and Tzeng, 2004). It is a distance-based method, based on an aggregation function indicating distance of each alternative to the ideal solution of the decision-making problem, and proposes compromise ranking, as well as one or as a set of compromise solutions that is the closest one to the ideal point. VIKOR utilizes a maximum group utility for the "majority" and a minimum individual regret for the "opponent" (Opricovic and Tzeng, 2007), which allows for specifying different ranking lists, according to different strategies, followed by decision-makers. The ranking of the alternatives may change, with respect to the weights of the criteria reflecting the preferences of decision-makers, according to the relative importance of the criteria. Opricovic (1998), Opricovic and Tzeng (2004, 2007), and Gul et al. (2016), can be referred to for more detailed information on the background of VIKOR, its stepwise implementation procedure, and application areas, including logistics and supply chain management, natural resources and environmental management, and energy management.

TOPSIS (Technique for Order Preference by Similarity to Ideal Solution) (Hwang and Yoon, 1981) is a method for comparison and ranking of alternatives, based on the principle of selecting the alternatives that concurrently have the shortest Euclidean distance from the positive ideal solution (that maximizes the benefits and minimizes the costs), and the farthest Euclidean distance from the negative ideal solution (that maximizes the costs and minimizes the benefits). These distances are calculated for each alternative and aggregated in a criterion that combines the two factors to be used as a base for ranking the alternatives. The values of attributes must be numeric and monotonically increasing or decreasing, to apply the TOPSIS technique, to locate the positive and negative ideal solutions, and to compare the Euclidean distances easily. For detailed information on the background of TOPSIS, its stepwise procedure and application areas, refer to Hwang and Yoon (1981) and Behzadian et al. (2012).

ELECTRE (ELimination Et Choix Traduisant la Réalité or Elimination and Choice Translating Reality) and PROMETHEE (Preference Ranking Organization Method for Enrichment Evaluations) are the most popular members of the family of outranking methods, of which basic concept is to deal with outranking relations, by using pairwise comparisons between alternatives under each criteria, separately. The first version of ELECTRE methods, ELECTRE I, is proposed by Roy (1968). After the first study on ELECTRE I, different versions of ELECTRE methods were introduced such as ELECTRE II, ELECTRE III, ELECTRE IV, ELECTRE IS, and ELECTRE TRI (renamed as ELECTRE TRI-B after the development of its new version, ELECTRE TRI-C and ELECTRE TRI-nC as an extension of ELECTRE TRI-C). Each version of the ELECTRE methods were developed for proposing solutions to various types of problematics such as choice problematic, ranking problematic, and sorting problematic (Figueira et al., 2010). ELECTRE I and IS were designed for choice problematic, which aims to identify the smallest set of alternatives for selecting the best among them. ELECTRE II, III, and IV were designed to be employed in the ranking problematic, which aims to rank the alternatives from the best to the worst. ELECTRE TRI, TRI-C, and TRI-nC were proposed for sorting problematics aiming, in the assignment of the alternatives to predefined categories. The calculation process of ELECTRE I method is based on pairwise comparison of the alternatives so as to build outranking relations. The concordance and discordance indices are utilized to analyze the outranking relations between the alternatives. The dominated alternatives are eliminated by considering threshold values, defined via dominance relationships, and the partial ranking of the alternatives is determined. PROMETHEE was introduced by Brans (1982), and further developed by Brans and Vincke (1985). As in the case of ELECTRE, PROMETHEE also has a number of versions that were developed to aid in more complex decision-making problems (Brans and Mareschal, 2005). Among them, the PROMETHEE I and the PROMETHEE II have been developed for the partial ranking and complete ranking of the alternatives, respectively. PROMETHEE III was proposed for ranking problems based on intervals, the PROMETHEE IV for complete or partial ranking of the alternatives under the case of continuous set of viable solutions, the PROMETHEE V for problems with segmentation constraints (Brans and Mareschal, 1992), the PROMETHEE VI for the representation of human brain (Brans and Mareschal, 1995), the PROMETHEE GDSS for group decision-making problems (Macharis et al., 1998), and the visual interactive module Geometrical Analysis for Interactive Aid for graphical representation (Mareschal and Brans, 1988; Brans and Mareschal, 1994). The last versions of PROMETHEE has been recently introduced by Figueira et al. (2004), namely the PROMETHEE TRI for dealing with sorting problems and the PROMETHEE CLUSTER for nominal classification. Among them, PROMETHEE II is one of the most commonly used version, which provides a complete ranking of the alternatives. By this approach, a pairwise comparison of alternatives is conducted, based on criteria to be maximized or minimized. The implementation of PROMETHEE II depends on the determination of the weights of criteria, the preference function to convert the difference between evaluations related to two alternatives into a preference degree, the value of an indifference threshold, and a strict preference threshold. For more detailed information regarding to ELECTRE and PROMETHEE methods, Govindan and Jepsen (2016), and Behzadian et al. (2010) can be referred to, respectively. ORESTE (Organization, Rangement Et Synthese De Donnes Relationnelles) method is another member of the family of outranking methods. ORESTE does not require the quantification of criteria weights, and an ordinal ranking is obtained, based on a global preference structure on alternatives.

The Simple Additive Weighting (SAW) (MacGrimmon, 1968), approach, also known as the scoring method or weighted linear combination, calculates the performance score of each alternative that represents the goal achievement under each criterion, based on the weighted sum of the evaluation ratings regarding to the alternative. Here, the weights represent the relative importance of each attribute. The performance scores are used to rank the alternatives and select the best one. SAW is a widely used MADM approach due to its a simplicity.

DEMATEL (Decision-making trial and evaluation laboratory) analyzes the possible interdependencies between the attributes and converts them into a structural model, based on decision matrices that categorize the relationships as the cause group and effect group. An impact relation map is also obtained as a result of the DEMATEL procedure that ranks the factors and reveals the critical factors of the system. DEMATEL is an effective approach to handle complex problems with interdependent relationships between factors. For detailed information on DEMATEL technique and its multiple step procedure refer to Si et al. (2018).

Data envelopment analysis (DEA) (Charnes et al., 1978) is a linear programming approach to be utilized to measure efficiency and assess the multiple decision-making units, based on a number of inputs and outputs. While being commonly used to assess the efficiency of service units (e.g., bank branches), the technical efficiency of several system alternatives can also be determined by the method. The implementation procedure begins with calculating the weights associated also with the inputs and outputs of the decision-making units, which will then be multiplied with the actual values of input/output data to determine the efficiency scores. DEA is a nonparametric approach in that cost, profit, or production functions are not estimated from the data.

Multiple attribute utility analysis (MAUA) (Keeney and Raiffa, 1976) utilizes the utility theory that describes the selection of a satisfactory solution as the maximization of satisfaction derived from its selection (Kahraman, 2008). In MAUA, the utility functions are determined for each attribute, and used to measure the utility values for each alternative and the selection of the best one among them.

Apart from the MADM approaches that have been commonly used in MCDM problems, a variety of techniques have been proposed to provide efficient solutions to the decision problems having different natures. COPRAS (Zavadskas et al., 1994), MOORA (Brauers and Zavadskas, 2006), ARAS (Turskis and Zavadskas, 2010), COPRAS-G, ARAS-F, ARAS-G, and MULTIMOORA (Brauers and Zavadskas, 2010; Turskis and Zavadskas, 2010), LINMAP (Srinivasan and Shocker, 1973), EVAMIX (Voogd, 1982), SWARA (Keršuliene et al., 2010), EDAS (Ghorabaee et al., 2015), WASPAS and NAIADE (Munda, 1998) can be listed among them.

A classical crisp MADM methodology assumes that the values of attributes corresponding to alternative and associated criteria weights are known precisely and can be represented by crisp numbers. However, in many real-world decision problems in biomass-based production chains, the data related to an attribute may not be defined crisply, due to unavailable, imprecise, or incomplete information, hesitation of decision-makers, and changing economic, technical, and environmental conditions. The traditional MADM methods cannot handle such problems adequately. The application of the fuzzy set theory in the field of MADM is an appropriate method, when the data related to attributes and their weights cannot be defined crisply, but only as fuzzy numbers. The application of fuzzy versions of the MADM methods (e.g., fuzzy AHP, fuzzy ELECTRE, fuzzy TOPSIS, etc.) would provide more realistic and effective results in these situations. However, fuzzy data are operationally more difficult to manipulate than crisp data, and they certainly increase the computational requirements, as well as the computation time, in particular during the process of ranking when searching for the preferred alternatives (Chen and Hwang, 1992).

6.2.2.2 *Multiple Objective Decision-Making*

MODM methods provide efficient tools in modeling and optimization of biomass-based production chains, when multiple objectives that represent different economic, environmental, and social aspects need to be handled simultaneously, to obtain sustainable design and operations. MODM methods are also useful to capture varying overall goals corresponding to different stakeholders and decision-makers in the supply chain. For example, profitability may be the major consideration for the producer company, whereas the labor conditions and environmental impacts may be more important to align with the legal regulations and restrictions.

The goals may also be conflicting and noncommensurate, which requires taking into account the trade-offs between objectives. In this case, there is not a single solution to optimize all objectives and a pareto optimal set of solutions are obtained. A pareto optimal set contains good compromise solutions in that one of the objectives cannot be improved without sacrificing another one. For example, to maximize the amount of bio-fuel produced in the conversion plant and hence the revenues from bio-fuel sales, more biomass feedstock should be transported from the supply sites to the conversion plant, which results in increased transportation costs and accordingly, overall costs.

MODM models can be in different forms, such as linear, nonlinear, dynamic, or stochastic. MODM models enable a better representation of the real case by including many factors in the decision-making process. However, ensuring consistency in the formulation of different objectives is a challenging issue in MODM. A number of techniques have been introduced to solve MODM problems. Among them, the weighted-sum method has been commonly applied for environmental management problems. By the weighted-sum method, a single utility function is obtained through merging the objective functions of the MODM problem into one objective function using a weighted sum. Here, the weights represent the relative importance of each objective according to the decision-maker. This method is preferred due to its simplicity, however, it has several disadvantages. Firstly, using the weighted sum method, only efficient extreme solutions can be generated. This leads to many runs, some of which may be redundant in the sense that there may be many combinations of weights that result in the same efficient extreme solution. Secondly, the scaling of the objective functions has a strong influence on the results. To overcome these shortcomings, the ε-constraint method (Haimes et al., 1971) is commonly utilized to obtain pareto optimal solutions revealing the potential trade-offs and compromises among them and allowing the decision-makers to select the solution that best suits his/her preferences. The ε-constraint method aims to optimize only one objective function (commonly, it may be the most preferred or primary one), and to limit the others by some allowable ranges. By this way, the MODM problem is transformed into a single-objective problem and can be solved easily.

Assume the following MOMP problem (Mavrotas, 2009):

$$\begin{aligned}&\max/\min \quad (f_1(x), f_2(x), \ldots, f_m(x))\\&\textit{subject to } x \in S\end{aligned}$$

where, x is the vector of decision variables, $f_1(x), f_2(x), \ldots, f_m(x)$ are the m objective functions and S is the feasible region.

The ε-constraint method is one of the objective functions optimized, using the other objective functions as constraints, and incorporating them in the constraint part of the model as shown below (Chankong and Haimes, 1983):

$$\begin{aligned}\max/\min \quad & f_1(x)\\ \textit{subject to} \quad & f_2(x) \geq \varepsilon_2 \ \textit{for} \max \ \textit{functions}\\ & f_3(x) \leq \varepsilon_3 \ \textit{for} \min \ \textit{functions}\\ & f_m(x) \geq \varepsilon_m\\ & x \in S\end{aligned}$$

By introducing the ranges, $\varepsilon_i, i = \{1, \ldots m\}$ to objective functions, the efficient solutions of the problem are obtained.

Goal programming is another technique to model and solve MODM problems. Goal programming can be seen as an extension of linear programming to handle multiple and conflicting objectives, which are represented by a goal or target value to be accomplished, by minimizing the unwanted deviations from this set of target values in an achievement function. Goal programming can be used, either to determine the optimum resource combination to achieve a set of desired objectives, or to specify the level of satisfaction of objectives achieved by using the available resources.

By goal programming, a specific quantified goal (the desired level of achievement of the objective) is established for each of the objectives and the model is solved to obtain a solution, to minimize the sum of deviations of the objective functions from their corresponding goals. To this aim, three types of goals can be set; (1) an upper, one-sided goal that limits the objective function by a value that the decision-maker does not want to exceed (for minimization objectives), (2) a lower, one-sided goal that limits the objective function by a value that the decision-maker does not want to fall under (for maximization objectives), and (3) a two-sided goal to set a specific target that the decision-maker does not want to deviate from on either side. In the first case, a decision variable (s_i^+) is assigned that represents the amount, by which the goal is numerically exceeded and in the second case, a decision variable (s_i^-) is assigned that represents the amount, by which the goal is numerically under. These decision variables are referred to as deviational variables. If failure to meet goal i occurs when the attained value of an objective is numerically larger than the desired value of goal i, then the term s_i^+ will appear in the objective function. If failure to meet goal i occurs when the attained value of an objective is numerically smaller than the desired value of goal i, then the term s_i^- will appear in the objective function. In addition, if the goal is desired to be met exactly (without any deviation), both s_i^+ and s_i^- will appear in the objective function to assign a penalty for going both over and under a goal (Winston, 1994).

Goal programming models can be preemptive or non-preemptive. Preemptive goal programming is a useful tool when the decision-maker may not be able to determine precisely the relative importance of the goals. To apply preemptive goal programming, goals must be ranked from the most important to the least important (Winston, 1994), so that the objectives of primary importance are initially optimized ignoring the lower priority objectives, and then those of secondary importance are optimized by holding the value of the first priority objective function constant. The process stops when all priorities are considered. Non-preemptive goal programming should be used when all objectives are of comparable importance so that direct comparisons of the objectives are possible. In this case, all the undesired deviations are multiplied by weights, reflecting their relative importance, and added together as a single sum in the same objective function, namely the achievement function. The subjective determination of weights may be a drawback of this approach, due to the different solutions yielded by different weights.

6.2.2.2.1 Fuzzy Multiple Objective Linear Programming

Multiple objective linear programming (MOLP) is one of the most popular MODM techniques to deal with complex decision problems with multiple and conflicting objectives. Formulation of an MOLP problem requires various parameters to be included in the formulation of the objective functions and the constraints, of which possible values may be assigned in an experimental manner, through estimations based on historical data or through subjective judgment of the experts. In each case, the parameters may be imprecise or incomplete due to a certain level of uncertainty. Hence, the integration of fuzzy set theory and MODM methods provides better tools to represent the imprecise and multiple objective nature of the decision- making problems in tandem.

The following model represents the general MOLP problem with K objectives:

$$\begin{aligned} &\text{maximize/minimize} && Z = Cx \\ &\textit{subject to} && Ax \leq b \\ & && x \geq 0 \end{aligned}$$

where, $Z=(z_1,z_2,\ldots,z_k)^T$ is the vector of objective functions, C is a $K\times N$ matrix of constants, x is an $N\times 1$ vector of the decision variables, A is an $M\times N$ matrix of constants, and b is an $M\times 1$ vector of constants.

In most real-world biomass-based system applications, input data of C, b, and A are usually fuzzy/imprecise because of incomplete, nonobtainable, or fluctuated information and are described in linguistic terms. For instance, capacity limit of energy production in a bio-energy facility (b) may be "around 1000" kW/year. Similarly, the unit sale price of electricity (C) may be expected to be "about 0.1" €/kWh, and estimates of technological coefficients (A) may be "around 120" kWh per ton of biomass. The fuzzy mathematical programming approach for handling the multiple objective problems was firstly introduced by Zimmermann (1978). He extended the fuzzy linear programming approach to the general MOLP problem with K objectives as given in following model:

$$\begin{gathered} Cx \succ Z \\ Ax \prec b \\ x \geq 0 \end{gathered}$$

He also suggests using the following linear membership function for each of the objective functions (the degree of membership of the goals) $(z_k(x), k=1,\ldots,K)$ of this problem.

$$\mu_k(z_k(x)) = \begin{cases} 1 & ;\ z_k(x) > z_k^{\max} \\ \dfrac{z_k(x) - z_k^{\min}}{z_k^{\max} - z_k^{\min}} & ;\ z_k^{\min} < z_k(x) \leq z_k^{\max} \\ 0 & ;\ z_k(x) \leq z_k^{\max} \end{cases}$$

where, $z_k^{\max}$ and $z_k^{\min}$ are aspiration levels for the objective functions (the values that the decision-maker wants to achieve for the objective functions).

Linear membership function for the ith constraint (the degree of membership of the constraints) $(\mu_i(Ax)_i)$ is defined as:

$$\mu_i(Ax)_i = \begin{cases} 1 & ;\ (Ax)_i \leq b_i \\ \dfrac{(b_i + d_i) - (Ax)_i}{d_i} & ;\ b_i < (Ax)_i \leq b_i + d_i \\ 0 & ;\ (Ax)_i > b_i + d_i \end{cases}$$

where, d_i is the limit of the allowed violation of the ith constraint.

When these linear membership functions are included in the fuzzy decision of Bellman and Zadeh (1970), the problem becomes choosing x, which satisfies the following equation:

$$max \min_{\substack{k=1\ldots K \\ i=1\ldots M}} \left\{\mu_k(z_k(x)), \mu_i(Ax)_i\right\}$$

By introducing the auxiliary variable λ, the following equivalent model can be obtained:

$$\begin{aligned} &\textit{maximize} && \lambda \\ &\textit{subject to} && \mu_k(z_k(x)) \geq \lambda \quad k = 1 \ldots K \\ &&& \mu_i(Ax)_i \geq \lambda \quad i = 1 \ldots M \\ &&& \lambda \geq 0 \quad x \geq 0 \end{aligned}$$

Integration of FST with MOLP provides further contributions besides reflecting uncertainties in the model parameters directly into the optimization processes. Firstly, the variation or vagueness of the decision-maker's aspiration level can be incorporated to the model and thus a more confident solution set can be generated. Also, the solution procedure of FMOLP is simplified, when compared with deterministic multiple objective programming as the FMOLP does not have to search for the satisfactory solution in a set of noninferior solutions by distance-based criteria, as required by the conventional solution procedure of multiple objective programming (Chang and Wang,1997). Various approaches have been proposed by researchers to derive a satisfactory solution for an MOLP problem-based on decision-makers' subjective judgments and preferences. Lai and Hwang (1994a) can be referred to for an extensive literature survey that systematically classifies FMODM approaches.

In the following, some of the fuzzy goal programming (FGP) and the fuzzy modeling approaches, namely Bellman and Zadeh's "min operator", Werner's "fuzzy and operator", and the weighed additive approach of Tiwari et al., are described. For detailed information about the other widely used fuzzy modeling approaches, such as Lin's Weighted Max-Min Approach, Li's Two-Phase Approach, Lai and Hwang's Approach, refer respectively to Lin (2004), Li (1990), Lai and Hwang (1992). Lai and Hwang (1994b) can also be referred to for information about other approaches.

Fuzzy Goal Programming In a standard goal programming formulation, the aspiration levels of goals and constraints are precisely defined. But in real-life situations the decision-maker may not sure about what attainments are precisely desired for each objective function. Furthermore, decision-makers may have a priority structure, so expect to achieve some goals before, or more than others. To avoid these drawbacks that may be faced in using goal programming, FGP is an appropriate approach, to handle imprecise aspiration levels or priorities in a fuzzy decision environment, and reflect relative flexibility of aspiration levels that are specified by the decision-maker. In FGP, membership function values of each objective are replaced by the deviational variables in standard goal programming formulation.

Narasimhan (1980) initially proposed FGP, extending the fuzzy multiple objective programming approach introduced by Zimmermann (1978).

A typical FGP problem formulation can be stated as to find $x_i, i = 1...n$ to satisfy:

$$Z_m(x_i) \prec \overline{Z}_m \quad m = 1...M$$
$$Z_k(x_i) \succ \overline{Z}_k \quad k = M+1...K$$
$$g_j(x_i) \le b_j \quad j = 1...J$$
$$x_i \ge 0 \quad i = 1...n$$

where, $Z_m(x_i)$ is the mth goal constraint, $Z_k(x_i)$ is the kth goal constraint, $\overline{Z}_m$ is the target value of the mth goal, $\overline{Z}_k$ is the target value of the kth goal, $g_j(x_i)$ is the jth inequality constraint, and b_j is the available resource of inequality constraint j. The symbols $\prec$ and $\succ$ express the fuzzified versions of $\le$ and $\ge$, which can be read as "approximately less/greater than, or equal to". "Approximately less than, or equal to" means, the mth goal is allowed to be spread to the right-hand-side of $\overline{Z}_m(\overline{Z}_m = l_m$ where l_m is the lower bound of the mth goal) with a certain range of r_m so that $\overline{Z}_m + r_m = u_m$ where u_m is the upper bound of the mth goal. Similarly, with "approximately greater than, or equal to", the kth goal is allowed to be spread to the left-hand-side of $\overline{Z}_k(\overline{Z}_k = u_k$ where u_k is the upper bound of the kth goal), with a certain range of p_k so that $\overline{Z}_k - p_k = l_k$ where l_k is the lower bound of the kth goal.

To establish the membership functions of objective functions in FGP, upper (u) and lower (l) limits have to be determined besides the aspiration levels for each goal by the decision-maker. Results, which are obtained by solving the single-objective linear programming models, considering one objective function each time, can be used as starting points to specify the upper and lower limits. Therefore, the feasibility of each fuzzy goal is guaranteed.

The pay-off table includes the maximum and minimum values of the results of each objective that is taken as the aspired level of achievement and the lowest acceptable level of achievement. Decision-makers can estimate of the upper and lower values for each goal using pay-off table (Table 6.1).

TABLE 6.1 The Pay-off Table

	Z_1	Z_2	...	Z_M
X^1	Z_{11}	Z_{12}	...	Z_{1M}
X^2	Z_{21}	Z_{22}	...	Z_{2M}
⋮	...	...	...	...
X^M	Z_{M1}	Z_{M2}	...	Z_{MM}

In Table 6.1 Z_m $(m = 1,\ldots,M)$ denotes the mth objective function, and X^m is the optimal solution of the mth single objective problem. Solving the problem with X^m for each objective, a pay-off matrix with entries $Z_{pm} = Z_m(X^p)$ $(m, p = 1,\ldots,M)$ can be formulated as presented in Table 6.1 upper and lower limits can be determined, respectively, as follows:

$$u_m = (Z_m)^{max} = max_p(Z_{pm}) \quad p = 1,2,\ldots,M$$
$$l_m = (Z_m)^{min} = min_p(Z_{pm}) \quad p = 1,2,\ldots,M$$
$$(Z_m)^{min} \le Z_m \le (Z_m)^{max}$$

After constructing upper and lower limits, the membership functions can be developed for each fuzzy goal and constraint as follows:

For "approximately less than, or equal to":

$$\mu_{Z_m}(x) = \begin{cases} 1 & ; Z_m(x) \le l_m \\ \dfrac{u_m - Z_m(x)}{u_m - l_m} & ; l_m < Z_m(x) \le u_m \\ 0 & ; Z_m(x) > u_m \end{cases}$$

For "approximately greater than or equal to";

$$\mu_{Z_k}(x) = \begin{cases} 1 & ; Z_k(x) > u_m \\ \dfrac{Z_k(x) - l_k}{u_k - l_k} & ; l_k < Z_k(x) \le u_k \\ 0 & ; Z_k(x) < l_k \end{cases}$$

The membership functions of b_j are assumed to be non-increasing linear functions between b_j and $b_j + d_j$. Linear membership function for the jth constraint (the degree of membership of the constraints) $(\mu_{g_j}(x))$ is defined as:

$$\mu_{g_j}(x) = \begin{cases} 1 & ; g_j(x) \le b_j \\ \dfrac{(b_j + d_j) - g_j(x)}{d_j} & ; b_j < g_j(x) \le b_j + d_j \\ 0 & ; g_j(x) > b_j + d_j \end{cases}$$

Using Bellman and Zadeh's (1970), approach, the optimum solution can be obtained maximizing the minimum degree of membership for the fuzzy decision:

$$\max_{x \in F} \mu_F(x) = \max_{x \in F} min\left[\mu_{Z_1}(x), \mu_{Z_2}(x), \ldots, \mu_{Z_k}(x)\right]$$

By introducing the auxiliary variable λ, the following equivalent linear programming problem can be obtained:

$$\begin{aligned} &\textit{maximize} && \lambda \\ &\textit{subject to} && \lambda \leq \mu_{Z_k} && k = 1 \ldots K \\ &&& \lambda \leq \mu_{g_j} && j = 1 \ldots J \\ &&& x_i \geq 0 && i = 1 \ldots n \\ &&& \lambda \in [0, 1] \end{aligned}$$

Bellman and Zadeh's "min operator": Zimmermann (1978), adapted Bellman and Zadeh's (1970), "min operator" approach to the MOLP model. This operator is usually applied due to its easy computation and several nice properties. However, the solution generated by min operator is not compensatory and does not guarantee pareto optimal (Topaloglu and Selim, 2010).

According to the *min operator* approach the decision problem is choosing x^* so that:

$$\mu(x^*) = max\ min\ [\mu_{z_1}(x), \mu_{z_2}(x), \ldots, \mu_{z_k}(x), \mu_{g_1}(x), \mu_{g_2}(x), \ldots, \mu_{g_k}(x)]$$

By introducing the auxiliary variable λ, the following conventional LP problem can be obtained:

$$\begin{aligned} &\textit{Maximize} && \lambda \\ &\textit{subject to} && \lambda \leq \mu_{Z_k} && k = 1 \ldots K \\ &&& \lambda \leq \mu_{g_j} && j = 1 \ldots J \\ &&& x_i \geq 0 && i = 1 \ldots n \\ &&& \lambda \in [0, 1] \\ &\textit{and other system constraints} \end{aligned}$$

Werners' "fuzzy and" operator: As Bellman and Zadeh's *"min operator"* focuses only on the maximization of the minimum membership grade, it is not a compensatory operator, which means that goals with a high degree of membership are not traded-off against goals with a low degree of membership. As with most of the decision problems in the real world, energy planning problems are neither non-compensatory, nor fully compensatory. Because of the non-compensatory nature of its aggregation operator, "min operator" is not very suitable for modeling real-world problems. For some computationally efficient compensatory operators, refer to Lai and Hwang (1994a).

Werners' *"fuzzy and"* operator is one of the most powerful compensatory operators that can be used in setting the objective function in fuzzy programming to investigate better results. Werners (1988) formulates the *"fuzzy and"* operator as follows:

$$\mu(x) = \gamma\ min\ \{\mu_{Z_k}(x)\} + (1 - \gamma)(1/K)\left(\sum_k \mu_{Z_k}(x)\right)$$

where, K is the total number of objectives, $\mu_k(x)$ is the membership function corresponding to kth objective, and γ is the coefficient of compensation defined within the interval [0,1]. By adopting "min operator", the following LP problem can be obtained:

$$\begin{array}{lll} \textit{Maximize} & \lambda + (1-\gamma)(1/K)\sum_k \lambda_k & \\ \textit{Subject to} & \mu_{Z_k}(x) \geq \lambda + \lambda_k, & k = 1\ldots.K, \\ & \mu_{g_j}(x) \geq \lambda & j = 1\ldots J \\ & x \geq 0 & \\ & \lambda, \gamma \in [0,1] & \end{array}$$

and other system constraints

Tiwari et al.'s weighed additive approach: To improve efficiency and reflect the relative importance of fuzzy goals, a weighted additive model is introduced by Hannan (1981). The basic concept is to use a single utility function to express the overall preference index, which may be any aggregated function of all achieved values of goals. We use the Tiwari et al. (1987) weighted additive approach to reflect the relative importance of fuzzy goals, which maximizes the weighted sum of the achievement levels of the fuzzy goals. Formulation of this approach can be stated as in the following:

$$\begin{array}{lll} \textit{Maximize} & \sum_i w_i \mu_{Z_i}(x) & \\ \textit{Subject to} & \mu_{Z_i}(x) \in [0,1] & \forall i \\ & x \geq 0 & \end{array}$$

and other system constraints

where w_i is the weight of the ith goal. There are some good fuzzy approaches in the literature to determine crisp weights of the goals in the fuzzy environment. Lai and Hwang (1994b), can be referred to for a detailed explanation of these approaches. In this dissertation, we assume that the bio-energy production firm is the dominant partner in the supply chain and makes the final decision about the weights of the goals.

6.3 DISCUSSION

Design and management of biomass-based production chains involve complex decision problems on feedstock supply, logistics and transportation operations, type, location and scale of storage, preprocessing and conversion plants, type and capacity of conversion/pretreatment processes and technologies, and type of products. Also, these problems often include incommensurable goals and environmental, economic, social, and technical objectives and criteria have to be considered simultaneously. The combination of different sustainability aspects impacts the technical, environmental, and economic success of a biomass utilization project.

The strategic decisions often focus on supporting the allocation of investment capital and the creation and analysis of policies/strategies in the biomass-based production sectors, whereas the tactical optimization models handle mid-term logistics, production planning, and scheduling problems. Decision models should integrate different planning levels and supply chain stages, especially for large integrated supply chains. Most published optimization studies in design, and management of biomass-based production chains, handle the decision problems from the producer organization's perspective, assuming that no conflict of interest exists between different stakeholders in the supply chain. This might work for the large integrated companies, however, for cases where different stakeholders have their own goals, the optimization procedure should consider different aspects and perspectives among independent actors.

MCDM can be integrated with GIS and Life Cycle Assessment (LCA) to aggregate the data related to geographical and environmental criteria into the analysis. GIS can be used for land suitability analyses to obtain a suitability map indicating the most appropriate sites (alternative sites) for development of a biomass-based production facility. The alternative locations for facilities can be defined to be used in MCDM according to population census, water bodies, county boundaries, and railway and road transportation networks. The environmental impacts of different alternative processes, bio-products, biomass feedstock and technologies can be analyzed and quantified by LCA. Afterwards, these data can be used in MCDM approach that handle environmental aspects in criteria set or in objective functions.

The classical MODM models generally assume that the parameters that form the objective function and aspiration levels regarding goals are expressed in crisp values. Similarly, in the traditional MADM approaches all values of the criteria and their degree of importance (i.e., relative weights) are presumed to be known precisely, which makes it possible to organize them in a crisp ranking. However, in many real-world decision problems, utilization of the traditional MODM and MADM methods may face practical issues that significantly impact the decisions, due to imprecision or vagueness in the data regarding to criteria set, objective function parameters, or goals. In many situations, the values of the attributes, parameters, or aspiration levels can only be expressed by linguistic terms or qualitatively, which certainly requires the use of a more suitable method to handle these problems. In these cases, the integration of FST and MCDM methods provide appropriate tools to capture inherent uncertainties in the decision problem.

REFERENCES

Behzadian, M., Kazemzadeh, R.B., Albadvi, A., Aghbasi, M., 2010. PROMETHEE: a comprehensive literature review on methodologies and applications. Eur. J. Oper. Res. 200, 198–215.

Behzadian, M., Otaghsara, S.K., Yazdani, M., Ignatius, J., 2012. A state-of the-art survey of TOPSIS applications. Expert Syst. Appl. 39 (17), 13051–13069.

Bellman, R.E., Zadeh, L.A., 1970. Decision-making in a fuzzy environment. Manag. Sci. 17, 141–164.

Brans, J.P., 1982. The Engineering of Decision: Elaboration Instruments of Decision Support Method. PROMETHEE. Laval University, Quebec, Canada.

Brans, J.P., Vincke, P., 1985. Note—a preference ranking organization method. Manag. Sci. 31 (6), 647–656.

Brans, J.P., Mareschal, B., 1992. Promethee V: MCDM problems with segmentation constraints,. Inf. Syst. Oper. Res. 30 (2), 85–96.

Brans, J.P., Mareschal, B., 1994. The PROMCALC & GAIA decision support system for multi-criteria decision aid,. Decis. Support Syst. 12 (4–5), 297–310.

Brans, J.P., Mareschal, B., 1995. The PROMETHEE VI procedure: how to differentiate hard from soft multi-criteria problems. J. Decis. Syst. 4 (3), 213–223.

Brans, J.P., Mareschal, B., 2005. Promethee methods, Multiple Criteria Decision Analysis: State of the Art Surveys, International Series in Operations Research & Management Science, 78. Springer, New York, NY.

Brauers, W.K., Zavadskas, E.K., 2006. The MOORA method and its application to privatization in a transition economy. Control Cybern. 35 (2), 445–469.

Brauers, W.K., Zavadskas, E.K., 2010. Project management by MULTIMOORA as an instrument for transition economies. Ukio Technologinis ir Ekonominis Vystymas 16 (1), 5–24.

Brown N., 2009, Ongoing case: solar energy PEIS, In: Case Digest: Section 106 in Action, Advisory Council on Historic Preservation, <www.achp.gov>.

Chambers, N., Griffiths, P., Lewis, K., Jenkin, N., 2004. Scotland's Footprint, A Resource Flow and Ecological Footprint Analysis in Scotland. Best Foot Forward Ltd, Oxford, UK.

Chang, N.B., Wang, S.F., 1997. A fuzzy goal programming approach for the optimal planning of metropolitan solid waste management system. Eur. J. Oper. Res. 99, 303–321.

Chankong, V., Haimes, Y.Y., 1983. Multi-objective Decision-Making: Theory and Methodology. North-Holland, New York, NY.

Charnes, A., Cooper, W.W., Rhodes, E., 1978. Measuring the efficiency of decision-making units,. Eur. J. Oper. Res. 2 (6), 429–444.

Chen, S.J., Hwang, C.L., 1992. Fuzzy multiple attribute decision-making methods. Fuzzy Multiple Attribute Decision-Making. Lecture Notes in Economics and Mathematical Systems. Springer, Berlin, Heidelberg, p. 375.

Chen, C.Z., Lin, Z.S., 2008. Multiple timescale analysis and factor analysis of energy ecological footprint growth in China 1953e2006. Energy Policy 36, 1666–1678.

Cucek, L., Klemes, J.J., Kravanja, Z., 2012. A review of footprint analysis tools for monitoring impacts on sustainability. J. Cleaner Prod. 34, 9–20.

De Benedetto, L., Klemeš, J., 2009. The environmental performance strategy map: an integrated LCA approach to support the strategic decision-making process. J. Cleaner Prod. 17 (10), 900–906.

Figueira J., de Smet Y., Brans J.P., 2004, MCDA methods for sorting and clustering problems: Promethee TRI and Promethee CLUSTER, Université Libre de Bruxelles, Service deMathématiques de la Gestion, Working Paper 2004/02. <http://www.ulb.ac.be/polytech/smg/indexpublications.htm>

Figueira, J.R., Greco, S., Roy, B., Słowiński, R., 2010. ELECTRE methods: Main features and recent developments. In: Zopounidis, C., Pardalos, P. (Eds.), Handbook of Multi-criteria Analysis. Applied Optimization. Springer, Berlin, Heidelberg, p. 103.

GFN (Global Footprint Network), 2009, Glossary, <www.footprintnetwork.org>.

GFN (Global Footprint Network), 2010, Footprint basics—overview, <www.footprintnetwork.org>.

Ghorabaee, M.K., Zavadskas, E.K., Olfat, L., Turskis, Z., 2015. Multi-criteria inventory classification using a new method of evaluation-based on distance from average solution (EDAS). Informatica 26 (3), 435–451.

Govindan, K., Jepsen, M.B., 2016. ELECTRE: a comprehensive literature review on methodologies and applications,. Eur. J. Oper. Res. 250 (1), 1–29.

Gul, M., Celik, E., Aydin, N., Taskin, A., Gumus, A., Guneri, F., 2016. A state of the art literature review of VIKOR and its fuzzy extensions on applications. Appl. Soft Comput. 46, 60–89.

Haimes, Y.Y., Lasdon, L.S., Wismer, D.A., 1971. On a bicriterion formulation of the problems of integrated system identification and system optimization. IEEE Trans. Syst. Man Cybern. 1 (3), 296–297.

Hannan, E.L., 1981. Some further comments on fuzzy priorities. Decis. Sci. 13, 337–339.

Hwang, C.L., Yoon, K., 1981. Multiple Attribute Decision-Making: Methods and Applications A State-of-the-Art Survey. Springer-Verlag, New York, NY.

Kahraman, C., 2008. Fuzzy Multi-criteria Decision-Making: Theory And Applications With Recent Developments. Springer, New York, NY.

Keeney, R.L., Raiffa, H., 1976. Decision Analysis With Multiple Conflicting Objectives. Wiley & Sons, New York, NY.

Keršuliene, V., Zavadskas, E.K., Turskis, Z., 2010. Selection of rational dispute resolution method by applying new step-wise weight assessment ratio analysis (SWARA). J. Bus. Econ. Manag. 11 (2), 243–258.

Kissinger, M., Gottlieb, D., 2010. Place oriented ecological footprint analysis—the case of Israel's grain supply. Ecol. Econ. 169 (8), 1639–1645.

Lai, Y.J., Hwang, C.L., 1992. Possibilistic linear programming for managing interest rate risk. Fuzzy Set. Syst. 49, 121–133.

Lai, Y.J., Hwang, C.L., 1994a. Fuzzy Multiple Objective Decision-Making: Methods and Applications. Springer, New York, NY.

Lai, Y.J., Hwang, C.L., 1994b. Fuzzy Multiple Objective Decision-Making: Methods and Applications. Springer-Verlag, Berlin.

Li, R.J., 1990. Multiple Objective Decision-Making in a Fuzzy Environment. PhD Thesis, Department of Industrial Engineering, Kansas State University, Manhattan, KS.

Lin, C.C., 2004. A weighted max-min model for fuzzy goal programming. Fuzzy Set. Syst. 142, 407–420.

MacGrimmon, K.R., 1968, Decision-making among multiple attribute alternatives: A survey and consolidated approach, The RAND Corporation, Memorandum No. RM-4823-ARPA.

Macharis, C., Brans, J.P., Mareschal, B., 1998. The GDSS PROMETHEE procedure—a PROMETHEE–GAIA-based procedure for group decision support. J. Decis. Syst. 7, 283–307.

Mareschal, B., Brans, J.P., 1988. Geometrical representations for MCDA,. Eur. J. Oper. Res. 34 (1), 69–77.

Mavrotas, G., 2009. Effective implementation of the e-constraint method in multi-objective mathematical Programming problems. Appl. Math. Comput. 213, 455–465.

Meisterling, K., Samaras, C., Schweizer, V., 2009. Decisions to reduce greenhouse gases from agriculture and product transport: LCA case study of organic and conventional wheat. J. Cleaner Prod. 17, 222–230.

Munda, G., 1998. A NAIADE-based approach for sustainability benchmarking. Int. J. Environ. Technol. Manag. 6 (1-2), 65–78.

Narasimhan, R., 1980. Goal programming in a fuzzy environment. Decis. Sci. 11, 325–336.

Opricovic, S., 1998. Multi-criteria optimization of civil engineering systems. Fac. Civ. Eng., Belgrade 2, 5–21.

Opricovic, S., Tzeng, G.H., 2004. Compromise solution by MCDM methods: a comparative analysis of VIKOR and TOPSIS. Eur. J. Oper. Res. 156 (2), 445–455.

Opricovic, S., Tzeng, G.H., 2007. Extended VIKOR method in comparison with outranking methods. Eur. J. Oper. Res. 178 (2), 514–529.

Palmer, P.A.R., 1998. Evaluating ecological footprints. Electron. Green J. 1 (9), <escholarship.org/uc/item/05k183c9>.

Roy, B., 1968. Classement et choix en presence de points de vue multiples (La methodeELECTRE). Revue Francaise D Informatique de Recherche Operationnelle 2 (8), 57–75.

Roy, B., 1996. Multi-criteria Methodology for Decision Aiding. Kluwer Academic Publishers, Dordrecht, The Netherlands.

Saaty, T., 1980. The Analytic Hierarchy Process. McGraw-Hill, New York, NY.

Saaty, T., 1996. Decision-making in Complex Environments: The Analytical Hierarchy Process for decision-Making with Dependence and Dependence and Feedback. RWS Publications, Pittsburgh, PA.

Santhanam N., 2011. Increasing India's Wind Energy Footprint, Expert Speak, PowerWatch INDIA, 66–69, <www.eai.in>.

Schindler, 2010. Energy and GHG footprint, A big step forward. <ccr.schindler.com>.

Si, S.-L., You, X.-Y., Liu, H.-C., Zhang, P., 2018. DEMATEL technique: A systematic review of the state-of-the-art literature on methodologies and applications. Math. Prob. Eng. 2018, Article ID 3696457.

Srinivasan, V., Shocker, A.D., 1973. Linear programming techniques for multi-dimensional analysis of preferences. Psychometrika 38, 337–369.

Stamford, L., Azapagic, A., 2011. Sustainability indicators for the assessment of nuclear power. Energy 36 (10), 6037–6057.

Stoeglehner, G., Levy, J.K., Neugebauer, G.C., 2005. Improving the ecological footprint of nuclear energy: a risk-based lifecycle assessment approach for critical infrastructure systems. Int. J. Crit. Infrastruct. 1 (4), 394–403.

Stoeglehner, G., Narodoslawsky, M., 2009. How sustainable are bio-fuels? Answers and further questions arising from an ecological footprint perspective. Bioresour. Technol. 100, 3825–3830.

Tiwari, R.N., Dharmar, S., Rao, J.R., 1987. Fuzzy goal programming: an additive method. Fuzzy Set. Syst. 24, 27–34.

Topaloglu, S., Selim, H., 2010. Nurse scheduling using fuzzy modeling approach. Fuzzy Set. Syst. 161, 1543–1563.

Turskis, Z., Zavadskas, E.K., 2010. A new fuzzy additive ratio assessment method (ARAS-F). Case study: the analysis of fuzzy multiple criteria in order to select the logistic centers location. Transport 25 (4), 423–432.

Voogd, H., 1982. Multi-criterion evaluation with mixed qualitative and quantitative data. Environ. Plan. Bull. 9, 221–236.

Werners, B., 1988. Aggregation models in mathematical programming. In: Mitra, G., Greenberg, H.J., Lootsma, F.A., Rijckaert, M.J., Zimmermann, H.-J. (Eds.), Mathematical Models for Decision Support. Springer-Verlag, Heidelberg, pp. 295–305.

Wiedmann, T., Minx, J., 2009. A definition of 'carbon footprint',. Ecol. Econ. Res. Trend. 1, 1–11.

Wiedmann, T., Barrett, J., 2011. A greenhouse gas footprint analysis of UK Central Government, 1990–2008,. Environ. Sci. Policy 14 (8), 1041–1051.

Winston, W., 1994. Operations Research Applications and Algorithms. Duxburg Press, California, CA.

Wright, L.A., Kemp, S., Williams, I., 2011. 'Carbon footprinting': towards a universally accepted definition. Carbon Manag. 2 (1), 61–72.

Yue, D., Kim, M.A., You, F., 2013. Design of sustainable product systems and supply chains with life cycle optimization-based on functional unit: general modeling framework, mixed-integer nonlinear programming algorithms and case study on hydrocarbon bio-fuels. ACS Sustain. Chem. Eng. 1, 1003–1014.

Zavadskas, E.K., Kaklauskas, A., Sarka, V., 1994. The new method of multicriteria complex proportional assessment of projects. Technol. Econ. Dev. Econ. 1, 131–139.

Zimmermann, H.J., 1978. Fuzzy programming and linear programming with several objective function. Fuzzy Set. Syst. 1, 45–55.

FURTHER READING

Opricovic, S., Tzeng, G.H., 2002. Multi-criteria planning of post-earthquake sustainable reconstruction. Comput-Aided Civ. Inf. Eng. 17 (3), 211–220.

Serafim O., 1998. Multi-criteria Optimization of Civil Engineering System, Faculty of Civil Engineering, Belgrade.

Zenonas, T., Kazimieras, Z.E., 2010. A novel method for multiple criteria analysis: Grey additive ratio assessment (ARAS-G) method. Informatica 21 (4), 597–610.

Chapter 7

Modeling and Optimization Approaches in Design and Management of Biomass-Based Production Chains

Chapter Outline

7.1 Mathematical Modeling-Based Approaches 185
- 7.1.1 Linear Programming 186
- 7.1.2 Nonlinear Programming 192
- 7.1.3 Methods and Tools to Solve Linear and Nonlinear Programming Problems 194

7.2 Simulation-Based Approaches 206
7.3 Artificial Intelligence-Based Approaches 213
- 7.3.1 Computational Search and Optimization Methods 213
- 7.3.2 Machine Learning 226

7.4 Geographic Information Systems 230
7.5 Game Theory 231
7.6 Discussion 234
References 235
Further Reading 236

7.1 MATHEMATICAL MODELING-BASED APPROACHES

Mathematical programming is one of the most well-known branches of operations research to model and solve optimization problems. The mathematical programming branch deals with the development of mathematical models that describe real-world phenomena and can be used to determine the optimal outcome of the optimization problem handled. Basically, an optimization problem represented by a mathematical programming model involves optimizing (maximizing or minimizing) a real function (the objective function), by systematically choosing the values of a number of variables from within an allowed set that is restricted by the set of constraints, and computing the value of the function. More generally, optimization includes determining the "best available" values of some objective function, provided that input data are defined. A variety of different types of objective functions and different types of inputs can be utilized to define the handled problem.

Decision-Making for Biomass-Based Production Chains. DOI: https://doi.org/10.1016/B978-0-12-814278-3.00007-8

A typical mathematical programming model is a set of equations consisting of objective function(s) and constraint(s), which are mathematically formulated, based on a set of decision variables and parameters. Here, the objective function is the mathematical representation of the optimization objective(s) (a function of the decision variables) specified by the decision-maker(s) to be maximized or minimized. The constraints are the limitations imposed by the technical restrictions of the system, and environmental, geographic, and/or economic conditions. The decision variables should completely define the decisions to be made. They generally describe the allocation of limited resources among competing decisions and activities, and the mathematical programming model aims at programming (or planning) the best possible course of resource allocation by determining the values of the decision variables. The main purpose of mathematical programming models, which are mathematical representations of real-world problems, is to optimize one or multiple function(s) comprising a number of decision variables under a set of constraints imposed. To do this, a set of input parameters is utilized, of which values are known precisely or imprecisely. The imprecise parameters can be defined within a range, and can be represented by probability distributions or by scenario construction.

Depending on the specific characteristics of the variables, objective function(s) and constraints, the mathematical programming models can be classified into four broad categories; linear programming (LP), integer programming (IP), mixed integer programming (MIP) and nonlinear programming (NLP) models. These categories and their usage in optimization of biomass-based production chains are explained in the following subsections.

7.1.1 Linear Programming

LP models are mathematical programming models dealing with the optimization of a linear objective function under linear equality and linear inequality constraints. The set of feasible solutions of an LP model can be defined as the intersection of finitely many half spaces restricted by linear inequalities. The aim of the LP algorithm is to find the optimum point that comprises the optimum values of a number of decision variables in this set, where the linear objective function has the maximum or minimum value. The standard form of typical maximization and minimization problems can be expressed as follows:

$$\begin{aligned}
&\textit{maximize}\; c^T x = c_1x_1 + c_2x_2 + \ldots + c_nx_n \\
&(\textit{objective function}) \\
&\textit{subject to}\;(\textit{constraints}) \\
&\left.\begin{aligned}
&a_{11}x_1 + a_{12}x_2 + \ldots + a_{1n}x_n \le b_1 \\
&a_{21}x_1 + a_{22}x_2 + \ldots + a_{2n}x_n \le b_2 \\
&\vdots \\
&a_{m1}x_1 + a_{m2}x_2 + \ldots + a_{mn}x_n \le b_m
\end{aligned}\right\} Ax \le b \\
&x_1 \ge 0, x_2 \ge 0, \ldots, x_n \ge 0\}\; x \ge 0
\end{aligned}$$

$$\begin{aligned}
&\textit{minimize}\; y^T b = y_1b_1 + y_2b_2 + \ldots + y_mb_m \\
&(\textit{objective function}) \\
&\textit{subject to}\;(\textit{constraints}) \\
&\left.\begin{aligned}
&y_1a_{11} + y_2a_{21} + \ldots + y_ma_{m1} \ge c_1 \\
&y_1a_{12} + y_2a_{22} + \ldots + y_ma_{m2} \ge c_2 \\
&\vdots \\
&y_1a_{1n} + y_2a_{2n} + \ldots + y_ma_{mn} \ge c_n
\end{aligned}\right\} y^TA \ge c^T \\
&y_1 \ge 0, y_2 \ge 0, \ldots, y_m \ge 0\}\; y \ge 0
\end{aligned}$$

where, *x/y* represents the vector of unknown decision variables to be determined, *c* and *b* are vectors of known coefficients (parameters) and *A* is a known matrix of coefficients (parameters). For more detailed information on the theoretical background of LP refer to Dantzig and Thapa (1997).

Integer Linear Programming (ILP) is a generalization of LP, in which some or all of the decision variables are constrained to be integers. The integer variables represent the quantities that can only be integer in real-world problems (e.g., the number of conversion plants to be constructed). There are three major variants of ILP; Pure Integer Linear Programming (Pure ILP), Mixed Integer Linear Programming (MILP), and 0−1 Integer Linear Programming (0−1 ILP). A mathematical model is called pure ILP model when all decision variables must be integers. MILP problems involve both discrete and continuous variables, in other words, some decision variables are restricted to be integers, whereas others are allowed to be nonintegers (e.g., the amount of biomass transported between supply site and storage facility in tons can be nonintegers, and the number of storage facilities to be constructed must be integer). In 0−1 ILP problems, the decision variables are constrained to be either 0 or 1 (e.g., in the decision of opening or not opening a blending facility, the opening decision is represented by 1, and the not opening decision is represented by 0).

ILP can be used in a wide range of application areas, including production planning, scheduling, transportation and logistics planning, supply chain configuration and network design. ILP-based optimization studies on design, modeling, and management of biomass-based production chains comprise all or some of these areas, depending on the aim and scope of the study. The modeling studies using ILP may comprise strategic, tactical, and operational decisions separately or simultaneously. The ILP models that capture strategic decisions are often investment intensive and focus on the configuration design of the supply chain network. The decisions, related to the selection of the location, production/treatment technology/process, and capacity/size as well as specifying the number of storage, pretreatment, and conversion facilities, are often handled by the strategic ILP models. Pure ILP and 0−1 ILP models are suitable for strategic decision-making.

However, there is an important interaction between the decisions related to the design and configuration of the production chain and the decisions related to biomass and bio-product flow (i.e., the allocation of supply/demand nodes to storage/conversion/pretreatment sites and the material amounts to be transported between these locations). Hence, the decision-makers often develop and use MILP models, so as to use both integer and noninteger variables to identify the optimal supply chain structure, together with determination of decisions related to optimal biomass and bio-product flow, and transport operations between supply chain nodes, which may be biomass source locations or collection sites, transshipment locations, pretreatment, storage, and/or conversion sites. These models may incorporate one or multiple economic, environmental, energetic, and/or social objective function(s) with regards to a constraint set involving capacity, supply/demand, flow conservation, and mass balance restrictions. In this standard MILP model, the location selection decision can be represented by, either binary variables that refer to whether or not a facility is constructed at a certain potential site, or integer variables that integrate the number and location decisions, that is, the number of facilities constructed at a certain potential site (can be zero if that site is not chosen for construction), and the decisions related to biomass/bio-product flow are represented by continuous variables, corresponding to the amount of material flowing from one node to another in the network (may be zero if no transportation between two nodes). The economic objective function mainly takes into account the capital investment costs (including the land and construction costs), and operational costs regarding to the facilities, which are correlated to the capacity/scale of the facility and conversion/treatment technology, to determine the facility location, and distance-related transportation costs and biomass purchasing costs to determine the optimal course of material flow. The biomass and bio-product flow decisions may also be affected by restrictions about supply and demand amounts. The decisions related to the capacity and scale of the facility and technology specifies the production amounts, which can be represented by continuous (noninteger) variables. For example, in the wood chipping case, the cost of an optimal arrangement of pretreatment facilities can be characterized by the chipping technology and chipping volume processed, as well as the optimized supply from the forest to the chipping facility. Capacity selection and capacity expansion decisions can be represented by either continuous (noninteger) or discrete (integer) variables. If the capacity of the main system or equipment that specifies the total capacity of the conversion/pretreatment/storage facility is chosen from a set of certain capacity levels, due to technical or technological restrictions, then binary variables can represent these decisions. Technology/process selection decisions are represented by binary variables that define whether or not to select a certain technology. Establishment of transportation links in a transportation network configuration is another strategic decision that can be represented by binary variables that define whether or not transportation links are established in a certain region.

Tactical and operational decisions in LP models are often represented by decision variables related to the short- to medium-term scheduling (determining amount, timing and routing) of biomass and bio-product production, storage, and shipment. These variables are generally noninteger, except the transportation mode decisions, which are represented by binary and integer variables that determine whether or not to use a particular mode, and the number of transportation vehicles to transfer the material on a particular route. The tactical and operational decisions are constrained by the established strategic decisions in the models that consider strategic and tactical/operational decisions in tandem. For comprehensive design and planning of biomass-based production chains, the optimization procedure should address simultaneous optimization of strategic decisions on the configuration design and tactical decisions on the operational schedules. However, some problems may require MILP models to optimize the periodical tactical/operational decisions for a given network structure. An example of pure tactical planning problems can be found in Gunnarsson et al. (2004), which studies the problem of deciding when and where forest residues have to be converted into chips, and how residues are to be transported and stored in order to satisfy the contracted demand at the saw mill. The MILP model involves both continuous and binary variables to determine the decisions, such as the biomass flows from harvest areas and saw mills to heating plants, whether forest residues are forwarded or chipped, whether a saw mill has been contracted, and whether a terminal is used in a certain time period under the restrictions of pre-established strategic decisions, such as chipping and storage capacity and saw mill and terminal locations.

The environmental objective function often comprises greenhouse gas (GHG) emissions associated with transportation and production/pretreatment processes. The amount of GHG emissions associated with transportation activities, which is highly related to the transportation distance has a significant impact on the supply/demand allocation and material flow decisions, while the GHG emissions, associated with production and pretreatment activities, affect the decisions related to number, capacity, and scaling of the facilities and technologies.

ILP models can either be single- or multiple-objective. As economic and environmental issues simultaneously affect strategic decisions, the multiobjective models represent the nature of biomass-based production chains in the decision-making process, more efficiently. Also, the multiple objective models can capture the trade-off between the revenues and investment/operational costs that higher costs, due to the larger plant/technology capacity/scale results in higher production amounts, and hence in higher revenues from product sales. A similar trade-off exists between the production and transportation-related GHG emissions in that the more bio-product/biomass is produced/transported, the higher the operational/transportation costs, but also higher revenues result from product sales.

The impact of time varying parameters, such as the temporal variability due to the seasonality of biomass supply and increasing bio-product demand on the tactical and operational decisions, is quite clear. However, strategic decisions are also influenced by these parameters. For example, the decision regarding the capacity of a conversion plant is highly correlated with the production amounts that a typical LP model selects. The capacity higher than the highest monthly production amount and the bio-product yield is dependent on the variability of biomass supply. To capture the impacts of the varying parameters related to biomass production and/or bio-product demand, these dynamics should be involved by multiperiod and/or multistage MILP that divides the annual cycle (i.e., time or planning horizon) into multiple time periods. In multiperiod MILP models, all tactical decisions and several strategic decisions can be made for each time period (e.g., facility open/not open decisions, terminal use/not use decisions, biomass flow decisions). However, different decisions related to the selection of location, technology, and scale for facilities for each time period may not be applicable, since changing a location, technology, or scale requires extensive costs once a facility is constructed. An example of a multiperiod MILP model that can be used for location, distribution, and production planning of anaerobic digestion-based biomass to energy supply chains is given in the following tables. The first table depicts the indices, parameters, and decision variables with corresponding descriptions, whereas the second table involves the objective function and constraints as well as related descriptions (Yılmaz Balaman, 2014).

Indices

p	Candidate locations for bio-energy plant construction
r	Biomass source sites
W	Waste feedstock types
E	Energy crop types
c	Plant capacity levels
t	Time periods

Parameters

PC_{pc}	Biomass capacity limit for the bio-energy plant at location p with capacity level c (t)
PEC_{pc}	Capacity limit for electrical energy production of the bio-energy plant at location p with capacity level c (kW)
e_W	Biomass to biogas conversion rate of waste biomass W (m^3/t)
e_E	Biomass to biogas conversion rate of energy crop E (m^3/t)
v_{be}	Biogas to electrical energy conversion rate (kWh/m^3)
c_{be}	Biogas to electrical energy conversion efficiency of the cogeneration unit (%)
TS_W	Total solid content of waste biomass W (%)
TS_E	Total solid content of energy crop E (%)
$MinTS$	Minimum total solid content of biomass slurry in digester (%)
$MaxTS$	Maximum total solid content of biomass slurry in digester (%)
AA_{rt}	Number of animals at biomass source site $r$at time period t
k_W	Monthly waste amount per animal (t)
VE_t	Energy crop cultivation amount at time period t (t/da)
EA_{rt}	Arable empty area in biomass source site r at time period t (da)
IC_{pc}	Discounted investment cost per kilowatt of installed capacity of the bio-energy plant at location p with capacity level c (€/kW_e)
OC_{ptc}	Discounted operational cost per kilowatt of installed capacity of the bio-energy plant at location p with capacity level c at time period t (€/kW)
TC_{rtpW}	Unit transportation cost of waste biomass W from biomass source site r to the bio-energy plant at location p at time period t (€/t)

TC_{rtpE} Unit transportation cost of energy crop *E* from biomass source site *r* to the bio-energy plant at location *p* at time period *t* (€/t)
EP Price of electrical energy (€/kWh)
WP Price of water (€/m^3)
CW_W Purchasing cost of waste biomass *W* (€/t)
CE_E Purchasing cost of energy crop *E* (€/t)

Decision Variables

Y_{pc} 0–1 integer variable that indicates whether or not a bio-energy plant with biomass processing capacity level *c* to be constructed in location *p*
BW_{rtpW} Amount of waste biomass *W* transported to the bio-energy plant at location *p* from the biomass source site *r* at time period *t* (t)
BE_{rtpE} Amount of energy crop *E* transported to the bio-energy plant at location *p* from the biomass source site *r*at time period *t* (t)
W_{pt} Water usage amount of bio-energy plants at location *p* (t)
OB_{pt} Amount of biogas output of bio-energy plant at location *p* at time period *t* (m^3)
OE_{pt} Amount of electrical energy output of bio-energy plant at location *p* at time period *t* (kWh)

The Objective Function: Maximization of the Total Supply Chain Profit

Total Profit = Revenue from Electricity Sale-(Total Discounted Investment Cost + Total Operational Cost + Total Transportation Cost + Biomass Purchasing Cost + Auxiliary Material Cost)

Formulations of the components of the Total Profit;

$$\textit{Revenue from Electricity Sale} = \mathrm{EP}^* \sum_p \sum_t OE_{pt}$$

$$\textit{Total Discounted Investment Cost} = \sum_p \sum_c IC_{pc}{}^*\mathrm{PEC}_{pc}{}^*Y_{pc}$$

$$\textit{Total Operational Cost} = \sum_p \sum_t \sum_c OC_{ptc}{}^*\mathrm{PEC}_{pc}{}^*Y_{pc}$$

$$\textit{Auxiliary Material}\ \text{(Water)}\ \textit{Cost} = \sum_p \sum_t W_{pt}{}^*\mathrm{WP}$$

$$\textit{Total Transportation Cost} = \left(\sum_r \sum_t \sum_p \sum_W TC_{rtpW}{}^*BW_{rtpW}\right) + \left(\sum_r \sum_t \sum_p \sum_{EC} TC_{rtpEC}{}^*BE_{rtpEC}\right)$$

(Continued)

(Continued)

$$\text{Biomass Purchasing Cost} = \left(\sum_r \sum_t \sum_d \sum_W CW_W{}^* BW_{rtdW}\right) + \left(\sum_r \sum_t \sum_d \sum_{EC} CE_{EC}{}^* BE_{rtdE}\right)$$

The Constraints

1. ***Biomass Supply Constraints***: To ensure the biomass procurement amount from a supply region must not exceed the total available biomass in that region.

$$\sum_p BW_{rtpW} \leq AA_{rtW}{}^* k_W \quad \forall r, \forall W, \forall t$$

$$\sum_p BE_{rtpE} \leq VE_t{}^* EA_{rt} \quad \forall r, \forall E, \forall t$$

2. ***Production and Capacity Constraints***: To limit the biomass processing capacity of the bio-energy plants to the maximum capacity of that plant. To calculate the biogas production and biogas to electricity conversion yield in the bio-energy plant at location p and ensure that the electrical energy production in the bio-energy plant at location p must not exceed the technical capacity limit for electrical energy production of the plant.

$$\sum_r \sum_W BW_{rtpW} + \sum_r \sum_E BE_{rtpE} \leq \sum_c PC_{pc}{}^* Y_{pc} \quad \forall p, \forall t$$

$$\left(\sum_r \sum_W BW_{rtpW}{}^* e_W\right) + \left(\sum_r \sum_E BE_{rtpE}{}^* e_E\right) = OB_{pt} \quad \forall p, \forall t$$

$$OB_{pt}{}^* v_{be}{}^* c_{be} = OE_{pt} \quad \forall p, \forall t$$

$$OE_{pt} \leq \sum_c PEC_{pc}{}^* Y_{pc} \quad \forall p, \forall t$$

3. ***Technical constraint***: To assure that the total solid content of the mixture of biomass slurry in the bio-energy plant must be between the technical limits to ensure the efficiency in the fermentation process in the digesters and to determine the amount of water required to satisfy the total solid content limits.

$$MinTS \leq \frac{\left(\sum_r \sum_W TS_W{}^* BW_{rtpW}\right) + \left(\sum_r \sum_E TS_E{}^* BE_{rtpE}\right)}{\sum_r \sum_W BW_{rtpW} + \sum_r \sum_E BE_{rtpE} + W_{pt}} \leq MaxTS \quad \forall p$$

7.1.2 Nonlinear Programming

A mathematical programming model is called an NLP problem if the objective function contains nonlinear function(s) and/or the feasible region is determined by the constraints involving nonlinear equalities or inequalities. By solving LP models, a global optimal solution (the best possible solution out of all possible solutions to a problem), a solution within the feasible region restricted by the set of constraints, with an objective value that is as good or better than all other feasible solutions to the model, can be obtained. However, this is not the case for nonlinear optimization models, which can behave much differently from LP models. NLP models may have a number of local optimal solutions, a solution in the feasible region, of which the immediate neighborhood (the area involving a solution set that is very similar, excluding a few insignificant details) contains no better feasible solutions than the given solution, and cannot be improved by considering a neighboring configuration. NLP solvers converge to a local optimal point, however, further local optimal points may exist outside the immediate neighborhood of the current solution, having better objective values in comparison with the current local optimum determined by the NLP solver. Metaheuristics are widely used to solve the NLP models to obtain a merely local optimal solution (or a good approximation to the global optimal solution).

A mathematical model is known to be nonlinear if it involves at least one equation/inequation that is not linear. A linear equality (or inequality) is a polynomial of degree one, and can take the form of $Ax + By + C = 0$, where x and y represent the decision variables. If any constraint or objective function of the problem cannot be mathematically represented in this form, then the model is said to be nonlinear. Higher degree polynomials, trigonometric functions, square, and cubic roots are nonlinear. Equalities/inequalities that contain multiplication or division of two or more decision variables are nonlinear. Quadratic programming is a special type of NLP, which involves maximizing or minimizing a quadratic function, consisting of several decision variables, subject to linear constraints on these variables. Fractional programming is another variant of NLP, of which its objective function is a ratio of two functions of decision variables. Objective functions with fractional values are often used to describe some kind of efficiency of a system.

The LP approaches provide useful methodologies to optimize biomass-based supply chains and logistics networks, however, some decision-making problems in design and management of biomass-based production chains involve cases that cannot be represented by linear formulations. For example, modeling and optimization of the production/conversion system and/or process itself require consideration of complex interactions and flows that have to be modeled with nonlinear formulations. Nonlinear mass balance, energy balance, exergy, time, and design constraints should be taken into account for different phases of the overall system/process (e.g., the drying, pyrolysis, oxidation and reduction phases for the gasification process), to optimize the production/conversion plant design in tandem with the supply chain. Another case that requires introducing nonlinear formulations in a MILP model is in addressing the decisions regarding conversion facility location and shipment routing, including the resulting traffic congestion impact simultaneously (Bai et al., 2011). In addition, nonlinear formulations may be needed to model the problems involving the case of energy content loss due to deterioration of the biomass overstayed in storage. In this case, the total energy content of the biomass, in storage in a time period, should be calculated as a function of the amount and energy content of the biomass remaining in storage from the previous time period, and the amount and energy content of the biomass purchased in the current time period, which are all decision variables that should be multiplied (Shabani and Sowlati, 2013). Another example of the cases that require the inclusion of nonlinear formulations, is the consideration of spatially correlated disruption possibilities in transportation links and intermodal hubs, especially those that occur with location-dependent probabilities, due to natural or man-made catastrophes such as hurricanes, flooding, and drought (Li and Ouyang, 2010; Marufuzzaman et al., 2014). In addition, formulating the minimization objective function involving transportation cost, based on the radius of the distribution area in location selection problems, may require using nonlinear equations.

7.1.3 Methods and Tools to Solve Linear and Nonlinear Programming Problems

There exists a range of methods to solve LP models. Simplex algorithms and interior-point methods (or barrier methods) are the most commonly employed techniques to solve LP models that are also used by commercial LP solvers. The branch and bound method and its variants are widely preferred by ILP and MILP solvers. Some solvers may also apply additional algorithms (e.g., decomposition methods) for large-scale problems, such as Column generation, Dantzig–Wolfe decomposition, and Benders' decomposition, as well as heuristic approaches.

7.1.3.1 Simplex Method and Interior Point Methods

The simplex algorithm was introduced by George B. Dantzig in 1947 (Dantzig, 1990). The algortihm provides a systematic procedure to solve LP problems, by producing a set of iterates by constructing a feasible solution at a vertex of the feasible search space (i.e., a convex region restricted by the linear constraints expressed as inequalities in the mathematical model containing possible values for decision variables), and iteratively moving along a path of edges of the search space, visiting and testing the adjacent vertex each time with a better value of objective function (a nondecreasing value for maximization objectives and a decreasing value for minimization objectives), until the optimum point is found, which is typically at one of the vertices.

The typical procedure of the simplex algorithm is explained step by step as follows:

Step 1. Transforming the LP model into the standard form.

The LP model has to be transformed into the standard form since the simplex algorithm operates on linear programs in a standard form, which is the basic format for all linear programs. This form is necessary because it provides an ideal starting point for solving the simplex method. The standard form has three main necessities; (1) it must have a maximization objective function, (2) all constraints have to be in a less-than-or-equal-to inequality, and (3) all variables have to be nonnegative. These requirements can be satisfied by transforming the given LP model (even minimization models), using basic algebra and substitution techniques. To transform an LP, with a minimization objective function into an LP with a maximization objective function, both the left and the right sides of the objective function are simply multiplied by -1. Linear constraints can be transformed from a greater-than-or-equal-to inequality to a less-than-or-equal-to inequality in a similar way as the objective function transformation, through multiplying both sides of the inequality by -1.

Step 2. Standard form of given LP problem is transformed into a canonical simplex tableau by introducing slack variables.

The inequality constraints are transformed into equality constraints by introducing slack variables, which will have a $+1$ coefficient in the standard form of an LP. A simplex tableau can be established after introducing the slack variables. This procedure is illustrated in Fig. 7.1, as an example. The canonical simplex tableau contains the coefficients corresponding to the objective function (in the last row) and the coefficients of the constraints (in the second and third rows). Row operations on the LP model are performed and a solution is checked for optimality using the tableau.

The model is checked for an optimal solution, which exists on the vertices of the feasible region, after construction of the tableau. If all

Standard form

$$
\begin{aligned}
&\textit{maximize} = 6x_1 + 8x_2 + 3x_3\\
&\textit{subject to}\\
&x_1 + 2x_2 + x_3 \le 12\\
&2x_1 + x_2 + 3x_3 \le 9\\
&x_1 \ge 0, x_2 \ge 0, x_3 \ge 0
\end{aligned}
$$

→

Standard form with slack variables

$$
\begin{aligned}
&\textit{maximize} = 6x_1 + 8x_2 + 3x_3\\
&\textit{subject to}\\
&x_1 + 2x_2 + x_3 + s_1 = 12\\
&2x_1 + x_2 + 3x_3 + s_2 = 9\\
&x_1 \ge 0, x_2 \ge 0, x_3 \ge 0, s_1 \ge 0, s_2 \ge 0
\end{aligned}
$$

x_1	x_2	x_3	s_1	s_2	z	b
1	2	1	1	0	0	12
2	1	3	0	1	0	9
–6	–8	–3	0	0	1	0

Canonical tableau

FIGURE 7.1 Simplex method.

values in the last row are greater-than-or-equal-to zero, it means that the variable has reached its optimal value. Otherwise, (if at least one value is less than zero) it means that optimality has not been achieved. The above table depicts that the optimal solution has not been reached yet, however a basic feasible solution, which satisfies the original set of equations can be produced.

Step 3. Identification of pivot variable and new tableau establishment.

Each step of the simplex procedure mainly manages the process of moving from a basic feasible solution to an adjacent basic feasible solution by a set of pivot operations. Hence, a sequence of improved basic feasible solutions is generated, with successively increasing objective function values, until the solution that maximizes the objective function is obtained. In each step, a pivot variable has to be selected, to establish a new tableau that contains basic feasible solution (an eventually the optimum solution), to continue the process.

The choice of the pivot variable at each step is based on the requirement that this pivot improves the objective value. The pivot variable is selected by looking at the last row of the tableau, the column above the smallest negative value in the bottom row (-8 in our example), contains the pivot variable (x_2). To find which variable will be pivot, the right-hand side values of the constraints (12 and 9) are divided by the corresponding values from the selected column ($12/2 = 6$ and $9/1 = 9$). The smallest nonnegative value (6 in our example), points out the pivot value (2 in our example).

After definition of the pivot variable and pivot value, the new simplex tableau can be constructed to find the new possible optimal solution. To this aim, the pivot variable is being optimized by a number of row operations while the rest of the tableau is kept the same. These row operations include:

1. The pivot variable is converted into a unit value (value of 1), by multiplying the row involving the pivot variable by the reciprocal of the pivot value. In the example above, the pivot variable is originally 2, so the entire row is multiplied by 1/2.
2. The other values in the column with the unit value have to be 0, since the pivot variable (x_2) in the second constraint is tried to be optimized. So, the value of pivot variable in other equalities are converted to 0 by applying row operations.
3. The row operations automatically make the values of variables in the columns and rows, other than the pivot column and row, recalculated in order to keep the tableau equivalent. This operation can be depicted by the following equation:
   New tableau value = [(Negative of the value in old tableau pivot column) × (value in new tableau pivot row)] + (Old tableau value)

Old simplex tableau:

Pivot value

x_1	x_2	x_3	s_1	s_2	z	b
1	(2)	1	1	0	0	12
2	1	3	0	1	0	9
-6	-8	-3	0	0	1	0

Old pivot column

Simplex tableau with unit value:

Unit value

x_1	x_2	x_3	s_1	s_2	z	b
1/2	(1)	1/2	1/2	0	0	6 ← New pivot row
	0					
	0					

Values to be made zero

New simplex tableau:

x_1	x_2	x_3	s_1	s_2	z	b
1/2	1	1/2	1/2	0	0	6
3/2	0	5/2	−1/2	1	0	3
−2	0	1	4	0	1	48

Step 4. Checking the optimality.

Optimality, which can be defined as finding the optimal solution with the maximum value of objective function, will need to be checked after establishing the new tableau. If all values in the last row are greater-than-or-equal-to zero, the solution is considered optimal. Otherwise, the procedure explained in Step 3 is repeated to identify a new pivot variable and construct the new simplex tableau. The procedure is repeated until the solution with maximum objective function value is achieved. The optimal values of decision variables can be identified once the tableau is proven to reach optimality. The optimality can be found by differentiating between the basic and nonbasic variables. A basic variable has a single value of 1 in its column with the remaining values being all zeros, otherwise the variable is nonbasic. The optimal solution of a nonbasic variable is zero, whereas the "*b*" value corresponding to the row that involves the value of 1 of a basic variable represents the optimal solution for that variable.

If the value of at least one of the basic variables is zero in the simplex tableau, this situation is called degeneracy, which may result in stalling occurring in practical problems, when solving by the simplex method. Stalling is the situation where no pivot variable can be found in succession to improve the objective function value, in other words, there will be no change in the solution even if the set of basic variables change. Cycling is another undesired situation in solving optimization problems by the simplex method. Cycling refers to the possibility of the same values of basic variables occurring subsequently twice, which means that the pivot operations of the simplex method will produce an infinite loop. In large-scale applications, degeneracy is common and stalling can be observed more commonly than cycling. New pivoting rules have been developed by researchers to avoid cycling.

In practical applications, the simplex method is quite efficient, and guarantees to find the global optimum if specific precautions are taken against cycling. The major drawback of the simplex method is that the number of iterations increase exponentially with the number of decision variables, which results in a time-consuming procedure with excess computational complexity, especially for medium- and large-scale problems. This drawback of the simplex method has led to a considerable research effort on algorithms with better computational complexity until interior-point methods with polynomial solution time have been developed.

Interior point methods are a family of algorithmic techniques developed to solve LP (and also NLP) problems in polynomial time and have received a considerable amount of attention from researchers, since Karmarkar (1984), developed a polynomial time LP algorithm, based on an interior point method, and proved its efficiency over the simplex method. The two methods have a few common strategies. However, interior point methods rapidly approach the optimal solution with fewer but relatively expensive iterations, whereas the simplex method comprises more in number but less expensive iterations to approach the optimal solution. Basically, interior point methods converge to a solution going through the middle (interior or exterior) of the feasible region defined by the constraints of the problem instead of the boundary (or around its surface). There exist two major interior point algorithms: the primal–dual interior point method and the barrier method. The primal–dual interior point method is a variant of the barrier method and is known to be a powerful technique to solve large-scale problems when high accuracy is needed. Both methods are based on Newton's method, which can be used to approach the optimal solutions within the feasible region, given a correctly formulated optimization problem. Newton's method iteratively attempts to find the roots of a differentiable function, in other words, the points that optimize an objective function. The method starts with a reasonable initial estimation and proceeds by approaching to its tangent line using calculus tools. Then the x-intercept of this tangent line is computed by elementary algebra, which will be a better approximation to the function's root, in comparison with the original estimation, and the method can be iterated. The number of iterations, to find the optimal solution by the primal–dual method, logarithmically increases with the number of decision variables.

7.1.3.2 Column Generation and Dantzig Wolfe Decomposition

Building and solving medium- and large-scale problems involve an important issue, so that the computational effort and computation time required to solve optimization problems, increases significantly with the number of variables, parameters, and constraints. To avoid the intractability, caused by using traditional approaches that handle all decisions in tandem through solving the entire problem, new approaches have been developed. One of these approaches is column generation, which is an efficient technique to deal with large-scale optimization problems. It solves a large-scale LP problem by reformulating it into two smaller size problems, a master problem and subproblems. Such a division leads to a great reduction in computational complexity and computation times, especially in too-large LP problems, to consider all the variables explicitly in the solution process. The basic idea of column generation is that a large number of the variables in the optimal solution will be nonbasic and take the value of 0, therefore the master problem is created that deals with a subset of variables with the potential to improve the objective function. The master problem is the original problem that handles a small, manageable part of the problem, whereas the subproblems are generated as new problems to capture new variables. To this aim, the variables are divided into two groups; the first group is handled by the master problem, of which elements are called generated columns, and the second group involves nongenerated columns. The division is illustrated in Fig. 7.2.

Column generation begins the process by solving a small part of the problem that is defined by the master problem and typically composed of several variables. After solving that part, the algorithm proceeds with analyzing the obtained partial solution, which is utilized in the objective function of the subproblem, to explore the next part of the problem that should be included in the solution process, to obtain an extended model. To this aim, the subproblem is solved, if the objective value of the subproblem is negative, a new variable with a negative reduced cost* is identified and added to the master problem. The extended model is then resolved. This procedure is

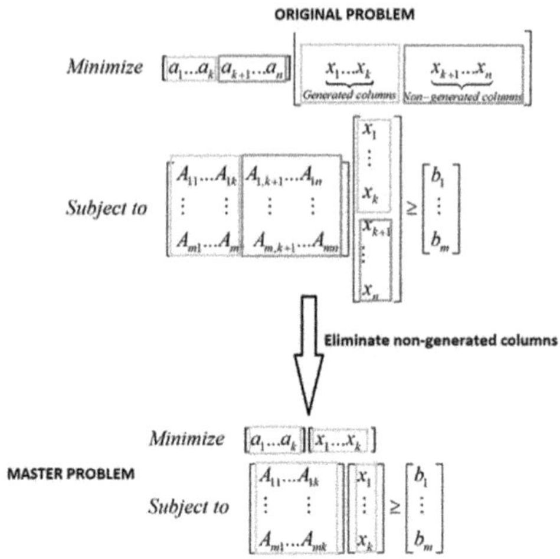

FIGURE 7.2 Column generation procedure.

repeated until a satisfactory solution to the overall problem is generated, or the subproblem with a nonnegative reduced cost is obtained, which means that the optimal solution to the master problem is found.

In summary, the basic idea behind the column generation technique is the separation of tasks in a large-scale linear optimization problem that had been previously considered as intractable. In other words, it separates the task of suggesting attractive alternatives for a decision problem from the task of selecting the best alternative in a small subset of alternatives. A particular method, based on the column generation technique, is the Dantzig–Wolfe decomposition algorithm, which is used to improve the tractability of large-scale LP problems. Starting with a feasible solution to the master problem, the Dantzig–Wolfe decomposition algorithm formulates new objective functions for each subproblem, utilizing the solution of the master problem in each step, with the aim of offering solutions to improve the current objective value of the master problem. Then, the subproblems are resolved using the reformulated objective function and the optimal solution is fed back to the master problem. The master problem involves one or more new variables (new columns), produced so far by subproblem solutions, based on those variables' individual capacity to improve the master problem's objective value. The number of iterations performed to solve the master problem is equal to the number of variables generated. The subproblems can be solved iteratively or in tandem, as they have completely independent solutions. In case of successive solutions, the master problem may incorporate all variables after waiting until the solutions to each of the subproblems have been obtained. A number of the commercially available mathematical modeling languages, such as AMPL and GAMS, involve the examples of implementation of the Dantzig–Wolfe decomposition algorithm.

* Reduced cost refers to the amount that the objective function coefficient of a variable would have to improve before it will be beneficial to give the corresponding variable a positive value in the optimal solution. For instance, if the reduced cost of a variable is five, that variable would be an attractive alternative to enter into the solution, only if the objective function coefficient associated with that variable decreases by five units in a minimization problem, and increases by five units in a maximization problem.

7.1.3.3 Cutting Plane Method and Benders' Decomposition

The cutting plane method is commonly used for solving ILP and MILP problems to find integer solutions, by solving the linear relaxation of the given integer programming model, which is a noninteger LP model. A cutting plane algorithm is generally used to search for valid inequalities that cut-off the noninteger solutions in two cases, when the set of constraints in our integer programming model is too large, and when the inequality constraints in the original integer programming model are not sufficient to yield an integer solution. The basic idea of the cutting plane method is iteratively refining the search region by introducing linear inequalities, known as cuts, maintaining the original feasible region. If the mathematical model is linear, an extreme or corner point in the feasible region can be the optimal solution, which may or may not be integer solution. If it is not integer, a linear inequality, in other words a cut, can be found to cut away a part of the feasible region, so as to separate the optimum solution. Therefore, a new separation problem is introduced and a cut is added to the relaxed LP model, which makes the existing noninteger solution no longer in the feasible region. This cutting process is repeated until the optimal solution found is also an integer solution. The steps of the cutting plane algorithm can be explained as follows:

1. The relaxed integer programming problem (the problem with continuous variables instead of discrete/integer variables) is solved.
2. Stop, if all variables in resulting solution have integer values, which means that it is the optimum.
3. Otherwise, generate a cut, that is, a constraint in the form of a linear inequality, which is satisfied by all feasible integer solutions.
4. Add this new constraint to the model, resolve the problem, and go back to step 2.

Benders' decomposition is a solution technique for solving large-scale mathematical programming problems. This technique is especially suitable for problems with a special block structure that generally exists in scenario-based stochastic programming applications. Benders' decomposition is based on a "divide-and-conquer" idea that the decision-making process is divided into several stages, by separating the variables in the original problem into a first-stage master problem with the first set of variables, and a second-stage subproblem with the second set of variables. The master problem is solved, and the solutions are used to determine the values for the second set of variables in the subproblem. If the first-stage decisions are found to be infeasible by the subproblem, one or more cuts, known as "Benders' cuts", are introduced into the master problem depending on the main idea of the cutting plane algorithm. The modified problem is then resolved and this procedure is repeated until no cuts can be created. The Benders' decomposition generates and adds new constraints into the master problem, while it proceeds towards the optimal solution, hence it uses row generation differently from Dantzig–Wolfe decomposition, which is based on column generation. By this way, a series of small-scale problems are solved, instead of a single large-scale problem, to overcome the computational complexity associated with solving larger problems.

Although only linear constraints are included in the master problem while Benders' decomposition procedure proceeds, the master problem does not have to be an LP problem. Instead, it can be a nonlinear, an integer, or a constraint programming problem. Hence, Benders' decomposition can be employed to solve a broader class of optimization problems. In some cases, more than one group of second-stage decisions can be made independently, so that multiple subproblems can be generated and solved individually to enable parallel implementation. This characteristic of Benders' decomposition is quite useful for decomposing and solving complex two-stage stochastic programming models, where some actions are taken in the first stage that are followed by random events represented by a set of scenarios. The first-stage decisions are impacted by the occurrence of random events, hence a recourse decision is made in the second stage to cope with the randomness caused by uncertainty. In these problems, the recourse problems in the second stage can be solved independently based on known first-stage decisions

7.1.3.4 Lagrangian Relaxation

Lagrangian relaxation is a relaxation and decomposition method to solve mathematical programming problems, which enables us to approach a difficult, constrained optimization problem by means of a simpler and relaxed problem. The solution to the relaxed problem can be accepted as an approximate solution to the original problem. The basic idea of Lagrangian relaxation is to separate the constraints into two groups, namely the easy constraints and the hard constraints, and eliminate the hard constraints from the constraint set of the mathematical programming model, by removing them to the objective function through a number of multipliers, to generate a relaxed version of the problem. Thus, the solution procedure is converted into solving a series of simplified problems involving the set of easy constraints, to derive better and better approximations to the optimum solution of the initial complex problem at each iteration. In this procedure, a multiplier called "Lagrange Multiplier" is used to impose a cost value for penalizing the violations of strict inequality constraints in the optimization model. These cost values are used instead of hard constraints in the relaxed problem, of which the optimal solution may, in some cases, actually be the same as the optimal solution of the original problem.

7.1.3.5 Branch and Bound

The branch and bound method is a "divide and conquer" algorithm for integer programming problems, which firstly divides a large-scale problem into several smaller ones (the branching part), and then estimates the goodness of a solution for each smaller problem (represented by nodes), according to the predetermined upper and lower bounds of the optimal solution (the bounding part). If necessary, the problem can further be divided until a smaller problem is obtained that can be handled easily. The algorithm searches the feasible solution space and systematically explores the candidate solutions, which are assumed to form a rooted tree with a set of candidate solutions at the root, by splitting the search space into smaller regions. The main aim of the algorithm is to enumerate the branches of this tree of all possible variable settings, each of which represents a subset of the solution set, after checking the upper and lower bounds of the branches (or regions of the search space). The branch is eliminated unless it produces a superior solution compared to the best one obtained so far by the algorithm. The efficiency of the algorithm mainly depends on the accurate estimation of the upper and lower bounds.

The generic steps of the algorithm for maximization problems can be given as follows:

1. Divide a problem into subproblems.
2. Find the optimal solution to the LP relaxation of a subproblem (the version of the existing integer programming problem with the integer restrictions relaxed).
3. Determine the upper and lower bounds at node 1 (the relaxed solution = the upper bound, the rounded down integer solution = the lower bound).
4. Establish two new nodes (i.e., subproblems) by selecting the decision variable with the greatest fractional part, and creating two new inequality constraints for this variable, one smaller-than-or-equal-to and one larger-than-or-equal-to.
5. Solve the LP relaxation of the model created by adding the constraint at each node.
6. Determine the lower and upper bounds at each node (the relaxed solution = the upper bound, the existing maximum integer solution obtained so far = the lower bound).
7. If the LP relaxation has an integer optimal solution with the greatest upper bound at any ending node, then the optimal integer solution is

reached. Otherwise, compare the solution of the LP relaxation (that is not all integer) with the best solution obtained so far (in other words the incumbent solution). If the optimal solution to the LP problem is worse than the incumbent solution, then stop, or else, return to step 4.

For minimization problems, LP relaxations are rounded up instead of rounding down and lower and upper bounds are reversed. Fig. 7.3 illustrates the branch and bound procedure.

There are variable selection rules (for branching procedure) and node selection rules (to determine which node to solve next) that specify the search procedure in the branch and bound algorithm. Choosing the variable with the greatest fractional part is common, however, strong branching may be another option for variable selection, which introduces tentative branching on each noninteger variable through a few iterations of the dual simplex, with the aim of testing which of the fractional variables provides the best improvement, before actual branching is performed on any of variables. Another branching strategy is pseudocost branching, which is based on keeping a memory of the success (the change in the objective function) of the variables already branched on. There are also a number of variations of these rules and hybrid rules can also be used that combine any two of these strategies. For detailed information on branching rules refer to Achterberg et al. (2005). Node selection can be performed by depth-first search, breath-first search, best-first search, or combinations of these strategies. In depth-first strategy, a node, with the highest level in the search tree (the deepest node), is chosen for exploration. In breath-first strategy, all nodes at a single level of the search tree are explored before jumping to any node at a higher level. The best-first strategy uses a priority queue that sorts the nodes, based on the upper or lower bounds obtained at each node, and the node with the best bound (highest upper bound in maximization) is chosen to explore.

There are several variants of the branch and bound method that are derived by utilizing decomposition methods, based on the idea of reducing the computational complexity and memory requirements when solving large-scale LP relaxations at each node. Among them, "branch and cut" is a method for solving ILP models, which is based on an integration of the branch and bound procedure and the cutting plane algorithm. The method basically applies branch and bound algorithm utilizing cutting planes to tighten and solve the LP relaxations at each node in the search tree. If the cutting plane algorithm is used only to solve the initial LP relaxation, the method is called cut and branch. Another variant is "branch and price", which combines the branch and bound method and the column generation algorithm. In the branch and price method, columns may be added to the LP relaxation at each node of the search tree.

There is a wide range of commercially available modeling languages and optimization software to support decision-making in linear and nonlinear

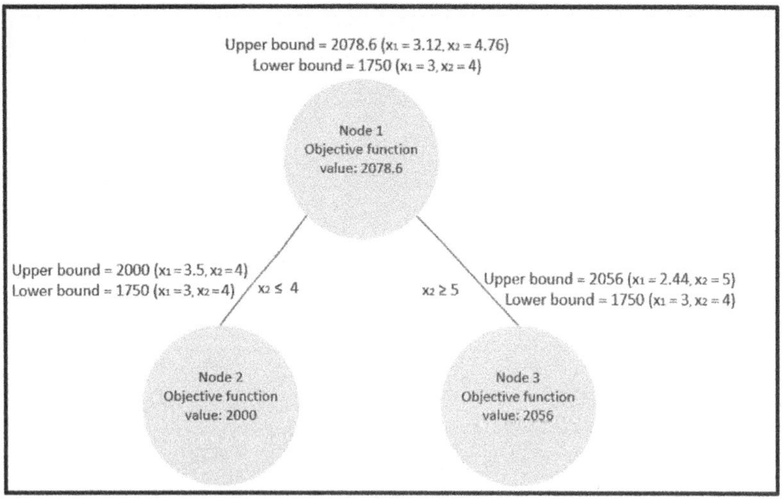

First branching

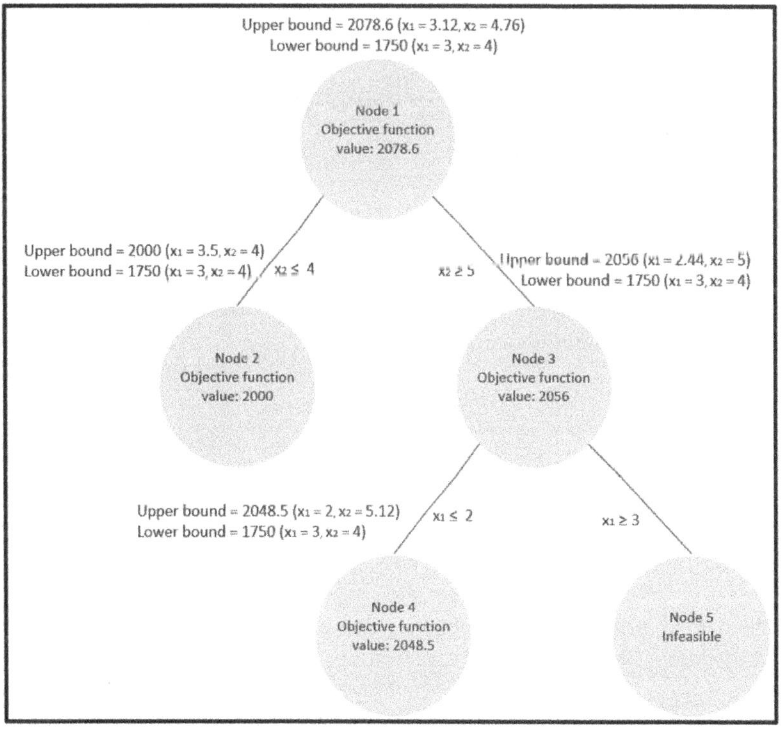

Second branching

FIGURE 7.3 Branch and bound procedure.

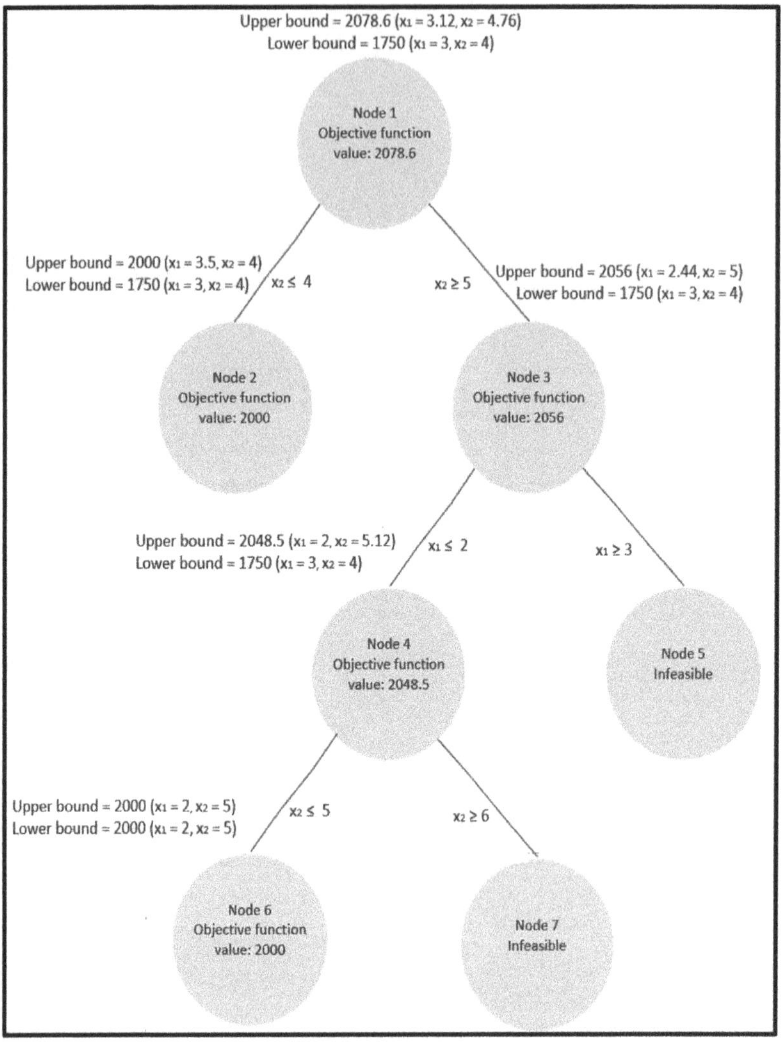

Third branching

FIGURE 7.3 Continued.

mathematical programming problems. Some examples of the most widely used modeling languages and software are provided in Table 7.1.

7.2 SIMULATION-BASED APPROACHES

Computer simulation models are developed and applied to represent, reproduce, and imitate the major characteristics, functions, operations, and behaviors of real-world systems, tasks, or processes over a certain time period.

TABLE 7.1 Some Examples of the Most Widely Used Modeling Languages and Software

	Linear Programming	Mixed-integer Linear Programming	Quadratic Programming	Nonlinear Programming	Mixed-Integer Nonlinear Programming	Stochastic Programming
AIMMS (Advanced interactive multidimensional modeling system)	✓	✓	✓	✓	✓	✓
AMPL (A mathematical programming language)	✓	✓	✓	✓	✓	
APMonitor (Advanced process monitor)	✓	✓	✓	✓	✓	
CPLEX (IBM ILOG CPLEX Optimization tudio) based on OPL (Optimization programming language)	✓	✓	✓			✓
FortMP	✓	✓	✓			
FortSP						✓
GAMS (General algebraic modeling system)	✓	✓		✓	✓	✓
	✓	✓	✓			
LINDO (Linear, interactive, and discrete optimizer)	✓	✓	✓	✓	✓	✓
Maple	✓	✓	✓	✓	✓	
MATLAB	✓	✓	✓	✓	✓	✓
Mathematica	✓	✓		✓	✓	
MOSEK	✓	✓	✓			
NMath	✓	✓	✓	✓	✓	
SNOPT	✓	✓		✓	✓	
WORHP				✓	✓	
XPRESS	✓	✓	✓	✓	✓	

Complex systems can be successfully modeled and studied by computer simulation, which is also a powerful tool to capture the randomness in the system parameters by probability distributions. Numeric models are generated in computer simulation to display and analyze the complex interactions between multiple variables within a system. Computer simulation tools and models are used in many areas including production planning, transportation and logistics planning, network design, performance optimization, project management, system and product design, economics, and finance. Through simulation approaches, the major impacts of key parameters, alternative system conditions, and different courses of action on the performance of the modeled system/process can be observed. The performance of the systems that are too complex to apply analytical methods can be estimated and new insights into new technologies can be explored by computer simulation. Simulation is useful, especially in the situations where carrying out experiments on the real system is costly, dangerous, or impossible (e.g., the system may be designed but may not be constructed yet). Simulation is also a useful tool for scientific modeling of natural systems.

The core stages in computer simulation basically comprise the development of the computer model of the real system, based on algorithms and mathematical equations, and the running of the model that contains these algorithms and equations. After the appropriate simulation model that best suits the system characteristics and requirements is selected, the model is built and implemented on the computer to visualize the simulation outcomes regarding the selected performance criteria. The critical issues in this process are obtaining accurate and valid data on the key characteristics and behaviors of the system, using proper approximations and assumptions to simplify the complex real-world system, and verification and validation of the simulation outcomes. Developing model validation and verification procedures are current and popular fields of research in computer simulation practices.

Another ongoing field of research in the scope of simulation is simulation–optimization, which integrates optimization methods with simulation analysis. The main aim of simulation–optimization is to find the optimal values for the system variables with minimum time and computation effort, of which impacts on the system behavior are analyzed considering different values by simulation experiment. This could also be conducted by running simulation experiments for all possible values of input variables, namely by parametric simulation methods, in which the input of each system variable is varied, leaving the values of other parameters constant, to observe the effect on the performance objective. However, this may be an intractable and time-consuming method, due to the complexity of the simulation model or the large number of possible values for input variables. By simulation–optimization, the optimal solution is obtained by running the simulation model to get a solution to the problem iteratively, and evaluating the obtained solution by an optimization procedure in each iteration, where the solution moves closer to the optimum solution. Here, several methods can be used to optimize the solution including statistical ranking and selection methods, response surface methodology, heuristic methods, and dynamic programming.

Computer simulation can be classified according to the attributes of the system and parameters, discrete and continuous, deterministic and stochastic, static or dynamic. Discrete event simulation (DES) is commonly used for computer simulation of engineering problems. This type of simulation manages the behavior of a complex system as a queue of events ordered, according to the simulated time that they should occur, and each event changes the system behavior at a specific point in time. DES generally used for countable phenomena to represent a system, which changes its behavior only in response to a finite number of events, discretely distributed over time. However, continuous simulation is applied to represent a continuously changing system. Continuous simulation solves differential equations for numerical analysis of the system and uses real numbers to change the state and output of the simulation. Chemical processes can be modeled using continuous simulation. Agent-based simulation is a variant of DES, in which the individual entities of the system (e.g., machines, operators, consumers, cells, etc.) are represented directly rather than by their attributes such as weight, operation hours, or density. Agent-based simulation is based on a set of rules that define the behaviors of the system and how the agent's state is changed from one time-step to the next.

Stochastic simulation models are used to model and analyze the systems composed of chance or random events that occur probabilistically, utilizing random number generators for probabilistic parameters. Monte Carlo simulation (MCS) is a commonly used stochastic simulation method in optimization studies. MCS applies iterative random sampling, using a large amount of random numbers, to obtain numerical results with regards to the performance of the system. After defining a domain of possible inputs, the input data is generated randomly, utilizing a probability distribution over the defined domain. The results are computed based on the generated input data and are eventually aggregated. Generally, MCS depends on the two random sampling processes, Latin hypercube sampling and simple random sampling. To obtain a good approximation of the optimal solution, MCS requires a large number of runs, however, its simplicity in application and property, so that its reliability and robustness are independent from the scale or complexity of the problem, makes it a powerful modeling technique.

Deterministic simulation is appropriate for the cases that no random variables are required to represent the system to be modeled. These models use known inputs and a single set of outputs is obtained. They are easier to build and interpret compared to stochastic simulation models. Dynamic simulation is used to model and analyze the behavior of a system that changes over time due to usually varying input values. Ordinary differential equations or partial differential equations are utilized to describe the systems in dynamic simulation. Dynamic simulation models can be run in real-time, as an online decision-support tool, to give a virtual response to the actual system on time, and hence early corrective or preventative actions can be taken for the hazards or problems. System dynamics is a simulation-based approach, which represents the nonlinear behavior of a complex system by a causal loop diagram. This diagram is in the form of a map that captures all components, interactions, internal feedback loops, and time delays in the system, and enables us to understand the structure of the system, to analyze its behavior over a particular time period.

The performance of design and management practices, in supply chains and individual companies in the biomass-based production sector, can be extensively improved by using computer simulation models and tools, which are able to monitor the real-time system performance and provide alternative plans when significant variations are observed in the initial program on production and logistics operations. Many studies developed conceptual computer simulation models, and used these models to represent the biomass-based production chain processes/systems, and evaluate a number of scenarios and solutions to design and management problems. The biomass-based production chains involve complex decisions and processes, usually requiring integrating long-term strategic design, medium/short-term resource allocation and routing, and mitigation of process disturbances due to material degradation. Computer simulation is a proper tool to handle this level of complexity and is especially suitable to support the tactical and operational level decision-making for the planning of logistics operations, including inventory management, production and transportation scheduling, and shipment routing.

Computer simulation is mostly used to model and analyze biomass/bio-product logistics and transportation systems. DES, which analyzes the changes in the state of the problem, based on the occurrence order of events at discrete time instants, is able to provide accurate simulation of logistics and transportation activities, to supply biomass sources to storage/pretreatment and conversion facilities, and to transfer bio-products to consumers. The appropriate number, type and capacity of the transportation vehicles needed to comply with the demand and the transportation costs can be determined by DES. In addition, allocation of resources, such as fuel, labor, equipment, and infrastructure for biomass and bio-product supply and transportation operations, can be determined maximizing the resource utilization. Various alternatives for biomass/bio-product transportation and logistics systems can also be compared according to their service level (i.e., supply time), cost, carbon emissions, and capacity characteristics. The specific attributes of two system/process options, such as efficiency, productivity, economic lifetime, investment, operations and maintenance costs, and energy consumption, to produce or treat a particular biomass/bio-product, can be compared by DES as well. Different biomass options can also be evaluated and compared with regards to the properties such as costs, moisture content, energy/fuel content, heating value, volume, and solid content to select the most appropriate biomass option or a mixture of biomass options, to produce a specific type of bio-product.

The economics of operating a biomass-based production system or supply chain can be evaluated, considering the fluctuations in different economic parameters, including labor and equipment costs, interest rates, unit feedstock and material costs, and unit transportation costs. To this aim, the computer simulation model can be run iteratively, each time considering a different combination of values corresponding to economic parameters and evaluating and comparing the results and the impacts of different parameter combinations on cost per unit of production. Computer simulation can also be utilized in operation scheduling for biomass and bio-product production, pretreatment, and transportation. The timing and order of such activities can be determined in order to maximize the resource and equipment utilization and minimize the costs. For example, harvest-time planning can be conducted taking into account varying ecological and geographical conditions.

The impacts of natural, climatic, biological, and geographic factors on delivery time and operational costs can be analyzed using stochastic and dynamic simulation models, utilizing the probability distributions obtained by analyzing the historical data. The simulation-based system dynamics approach can also be used to model and analyze the behavior of a multistage production chain, in the possibility of unavailability of supply chain nodes for certain durations, due to possible disruptions affecting different stages of the supply chain, at varying times and extents, over a certain time period. The results of the simulation provide some reconfiguration and recovery policies by reorganizing the material flow during and after the disruption, considering the devastating impact of the disruption and reconfiguration costs (Ivanov et al., 2010).

The complex behavioral assumptions regarding stakeholders of the supply chain can be captured by agent-based simulation. The social, socioeconomic, and environmental problems involving quality, energy and food security, and land-use issues can be handled using multiagent simulation models considering the interactions between stakeholders, risks associated with producers, and technological failures. A number of simulation models have been developed, in the area of design and management of biomass-based production chains, using commercial simulation tools, software, and languages (Chapter 7 can be referred to for information on simulation tools, software, and languages). Some examples include IBSAL (Sokhansanj et al., 2006), SHAM (Nilsson, 1999a and 1999b) and Biologics (Mol et al., 1997). Simulation methods can also be integrated with optimization techniques to overcome the shortcoming of pure optimization models, where they may not be capable of capturing all of the uncertain parameters and system dynamics. This integration is named as simulation optimization.

There are variety of commercially available simulations tools and software. Some examples are given in Table 7.2 along with corresponding brief descriptions.

TABLE 7.2 Some Examples of the Commercially Available Simulation Tools and Software

Name of the Software	Brief Description
ADINA	Engineering applications involving fluid-flow and heat transfer
ANSYS	Engineering applications
AnyLogic	Business and science applications
Arena	Discrete event simulation based on flowchart modeling
Chemical WorkBench	Chemical process simulation
COMSOL Multiphysics	Physics and engineering applications
DWSIM	Chemical process simulation
Ecolego	Simulation of dynamic models for deterministic and probabilistic cases
ExtendSim	Discrete event and agent-based simulation
Flexsim	Discrete event simulation
Fluent	For industrial applications involving fluid -low, heat transfer, reactions.
Khimera	Chemical process simulation
MATLAB	Programming, modeling, and simulation of a wide range of engineering applications
Plant Simulation	Simulation of industrial plants, lines and processes
PRO/II	Chemical process simulation, especially used for oil and gas production processes.
Simcad Pro	Dynamic discrete and continuous simulation
Simio	Discrete event and agent-based simulation
Simulink	Simulation based on block diagrams for electrical mechanical systems and machines
SRM Engine Suite	Simulation of engineering and energy applications including fuel production, combustion, and gas emissions.
TRNSYS	Simulation of energy applications including renewable energy systems and use of energy in buildings

7.3 ARTIFICIAL INTELLIGENCE-BASED APPROACHES

Artificial intelligence (AI) basically defines the ability of a digital computer to solve complicated problems and make decisions, by trying to understand the thinking of human brain and actions associated with intelligent beings. AI-based systems and techniques utilize the intellectual processes mainly performed by human brain such as reasoning, knowledge representation, systematic search and decision-making, language processing, meaning, learning from past experience, perception, and generalizing. Developing AI-based decision-making procedures has gained great advances in recent years in a wide range of application areas, including planning and optimization, intelligent search, image understanding and processing, big data and data mining, pattern recognition, and robotics. A wide range of methods and tools are used in the scope of AI, which can be categorized as computational search and optimization methods, logic-based methods (i.e., fuzzy logic), neural networks and methods based on statistics and probability.

7.3.1 Computational Search and Optimization Methods

The solution to many optimization problems can be obtained, through intelligently searching among multiple candidate solutions in a similar way to reasoning, characteristic of the human brain. Computational search algorithms for planning problems try to find a best-guess for the path, on which the solution lies, to the target goal, applying a set of inference rules in each step iteratively. Each method applies its own inference rules characterized by the characteristics of the associated intelligent beings. Simple exhaustive search methods are generally too slow and may be endless, especially when the search space is too large, which is usually the case for most large-scale, real-world problems. Heuristic approaches use specific algorithms to provide useful and efficient tools to handle such situations. Heuristic algorithms prioritize solution options in favor of those that are more likely to lead to the goal, to reduce the number of steps to reach the optimal solution. Some algorithms completely eliminate the options that are most unlikely to reach the objective so that the search space for solutions can be restricted into a smaller sample size.

Heuristic methods employ practical algorithmic procedures that can be defined as mental shortcuts to facilitate the decision-making process with cognitive load. Heuristics do not guarantee the optimal or best solution, but enables the finding of a sufficiently good or satisfactory solution for the immediate targets. Heuristic algorithms are extensively used where finding an optimal solution is impossible or impractical, due to the high level of complexity or nonlinear nature of the problem, to speed up the problem-solving process. Heuristics provide mental shortcuts that ease the cognitive load of making a decision by sampling a set of solutions in the large-scale problems where complete sampling is too time-consuming.

For many cases, the heuristic algorithm begins the search with an arbitrary solution to a problem, and continues the process by incrementally changing an element of the solution, to improve the result in terms of the objective function value until no more improvements can be achieved. These algorithms are known as *hill climbing*, which is a mathematical optimization technique beginning the search at a random point in the feasible solution set, and attempts to find a better solution by keep moving the guess uphill, until we the top is reached. Hill climbing is a simple algorithm and produces good solutions in a limited time or in cases where the global optimal solution, or a close approximation to the global optimal solution, can be converged by a small number of increments, however, it belongs to the family of local search, which means it cannot guarantee to lead to an existing better solution, and generally cannot achieve the global optimum solution, but it is able to just find a local optimum. To continue the search without being restricted by local optima, several approaches have been developed. For example, the procedure may be restarted using a repeated local search, more complex iterative local search algorithms may be used such as an iterated local search. Also, more advanced algorithms have been developed, to use the neighbors of a solution to reach the best approximation of the global optimum, by exploring a larger search space. These algorithms are called *metaheuristics*, which prefer better neighbors to derive new solutions that improve the objective function, although they also accept worse neighbors in order not to get stuck in local optima.

A metaheuristic is known to be a high level algorithmic procedure aimed at finding a sufficiently good solution to optimization problems, especially with nonlinear objective function(s) or constraints, incomplete information, or limited computation time. Metaheuristics require less computational effort, in comparison with simple heuristics, simple exhaustive search, iterative algorithms, and optimization methods, in solving large-scale mathematical optimization problems. Some common characteristics of metaheuristics are that they provide rule-based strategies for leading the process of search within the feasible solution space to efficiently explore the approximately optimal solutions. They utilize several assumptions, which make them applicable for a wide range of problems, hence they are not problem specific. Some algorithms involve complex learning processes to converge global optimum, whereas some of them are local search procedures. A number of metaheuristic approaches have been developed to improve the local search procedure of hill climbing including tabu search (TS), simulated annealing (SA), and greedy randomized adaptive search (GRASP), which can be both local search-based or global-search algorithms. Other types of metaheuristics are defined as population-based metaheuristics [e.g., evolutionary algorithms (EAs), including well-known genetic algorithms (GAs), particle swarm optimization (PSO), artificial bee colony, and ant colony optimization (ACO)]. Population-based metaheuristics use algorithms and rules, based on population characteristics, to guide the search within the solution space and improve a number of candidate solutions. Some of the population-based metaheuristics depend on the collective behavior of agents in a population and are called swarm intelligence. Some examples to this category include PSO, artificial bee colony, and ACO. Another classification is nature-inspired metaheuristics, of which search algorithms and rules are inspired by the behavior of natural systems, including evolutionary computation-based algorithms, SA, ACO, artificial bee colony, and PSO. Metaheuristic algorithms can be used in combination with other optimization and statistical methods, such as mathematical programming, stochastic optimization, k-means clustering, and machine learning. For example, a metaheuristic can select an arbitrary solution in the feasible solution space restricted by the constraints of a mathematical programming model as the initial solution, and after calculating the associated objective function value the metaheuristic may choose to change the solution and try another one, based on its algorithm and specific search rules. The procedure continues until no improvements in the objective function value is achieved and the best approximate to the global optimum is reached. Theoretically, if metaheuristics are run for long enough, they can even find the global optimum solution. In the following subsections, the major metaheuristics that have been widely used, and that have potential to be used in modeling and optimization for design and management of biomass-based production chains, are summarized.

7.3.1.1 Greedy Algorithms

Greedy algorithms employ a problem-solving procedure to progressively build candidate solutions, to approximate the global optimum, by obtaining better and better locally optimal solutions at each stage. In general, greedy algorithms cannot yield a global optimal solution, but they may produce good locally optimal solutions in a reasonable time and with less computational effort. GRASP (Feo and Resende, 1989), is a well-known iterative local search-based greedy algorithm that involves a number of iterations to construct greedy randomized solutions and improve them successively. The algorithm consists of two main stages, construction and local search, to initially construct a solution, and then repair this solution to achieve feasibility. The algorithm produces greedy randomized solutions, by selecting new elements from a candidate set of greedy solutions constructed, based on the level of improvement on the partial solution under construction, using a greedy evaluation function. The selected elements are then incorporated into the current partial solution without destroying feasibility (if feasibility is destroyed, a new element is selected), until a complete feasible solution, of which neighborhood is investigated until a local optimum found by the local search procedure, is obtained. A list, namely a restricted candidate list, with the best elements, is formed by the evaluation of the elements by the greedy evaluation function. Then an element is randomly chosen from the restricted candidate list to incorporate it into the partial solution, and once the new element is included into the partial solution, the candidate set of greedy solutions is updated along with the greedy evaluation function.

7.3.1.2 Simulated Annealing

SA is a metaheuristic to approximate the global optimum of a given objective function by searching in a large solution space. SA is commonly used when the search space is composed of discrete (integer) variables. The search procedure is based on the probability that at each step, the algorithm probabilistically decides whether to evaluate the candidate solution in the neighborhood of the current solution, or to stay in the current solution. SA may be useful when finding an approximate global optimal is preferred to finding an exact local optimum.

SA algorithm is inspired by the annealing in metallurgy and materials science, which is a treatment technique, to change the physical and chemical characteristics of a material, to obtain a more durable and workable one. The technique involves exposing the material heating above its recrystallization temperature and then by controlled slow cooling. Heating increases the internal energy of the system and allows the atoms to move freely, and when the system is cooled, a new configuration is discovered with lower energy. From the beginning of the heating process to the end of the cooling process different values for internal energy are exploited. Here, each solution is represented by a different internal energy of the system, and the SA algorithm searches for the approximate global optimal solution (the optimum value of the internal energy of the system to obtain the highest level of durability), from the beginning of heating (with the maximum temperature value), to the end of the cooling (with the minimum temperature value). Heating leads to free movement of the atoms and represents the relaxation of the acceptance criteria of the candidate solutions taken from the search space to be evaluated, whereas cooling represents narrowing the search space and focusing on the improvements. Once the system has cooled and reaches the optimum temperature, the internal energy at that temperature represents the point of, or close to, global optimum.

SA is widely used in design and optimization studies due to its simplicity in implementation. The algorithm is similar to that of hill climbing, the only difference is that the SA algorithm picks a candidate solution to be evaluated randomly, instead of picking the best candidate solution. If the selected point improves the objective value, then it is accepted, otherwise the algorithm selects a candidate solution with a certain probability. This probability value is analogous to the temperature in the natural annealing process. The probability value starts high and at a higher value of probability (higher temperatures—heating) the algorithm allows uphill moves (free move of atoms). It gradually decreases and at lower values of probability (lower temperatures—cooling) the moves become more unlikely (forming of new configuration). The standard procedure of the SA algorithm is illustrated in Fig. 7.4.

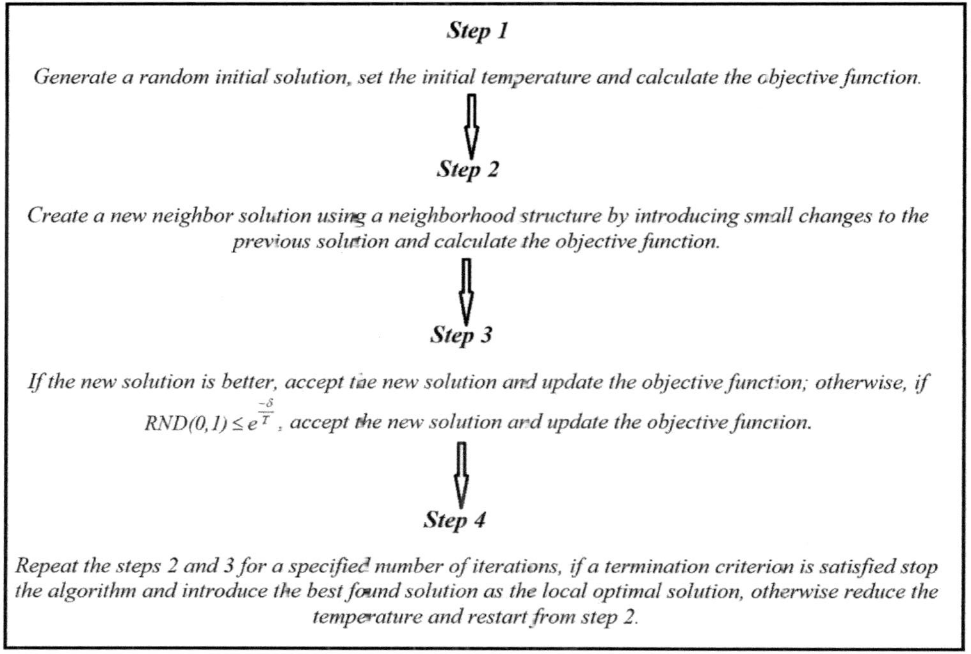

FIGURE 7.4 The standard procedure of the SA algorithm.

It should be noted that the function *RND (0,1)* produces random numbers between 0 and 1. The acceptance function—$e^{\frac{-\delta}{T}}$ (where T is the temperature value)—implies that there is more chance for small increases in the objective function to be accepted than large changes (since the solutions in the immediate neighborhood are searched by the local search procedure), and worse solutions are more likely to be accepted at higher temperatures.

7.3.1.3 Tabu Search

TS is based on local search methods that use immediate neighbors of a potential solution to search for a better solution, by moving from one potential solution to another, until a stopping criterion (a time limit or a threshold value) is satisfied. Local search procedures generally have the drawback of becoming stuck in local optima, that is, the poor-scoring areas with suboptimal and equally fit solutions.

TS is developed to mitigate the possibility of becoming stuck in a local optimum and to explore other regions in the search space, introducing a basic rule; firstly, worse moves are accepted if the search is stuck at a local optimum and no move improves the objective value and then, a tabu list (a list of prohibited moves) is introduced to prevent reevaluating the previously evaluated solutions. Hence, for carefully discovering the neighborhood of each solution, memory structures are established that describe the preevaluated solutions (i.e., forbidden solutions) and a set of other rules is provided by the decision-maker, and contains information related to the exploration process. Using these memory structures that involve the story of the search process, the solutions that have been previously visited within a certain time period, and the solutions that have violated a rule, are placed in the tabu list to prevent the algorithm to consider these solutions repeatedly. Basically, the tabu list involves the solutions that have been visited in recent past, in other words, a short-term set of solutions. Here, the number of previous iterations define the number of solutions to be stored that should not be revisited, which is called "tabu tenure".

The memory structures used in TS can be investigated under three categories; short-, intermediate-, and long-term memories. These memories can overlap in practice, however, the impact and frequency of the changes made can be considered to differentiate them. Short-term memory may be sufficient to find a solution better than the solutions obtained by traditional local search methods, however solving harder and large-scale problems may require intermediate- and long-term structures. A short-term structure involves the list of recently evaluated solutions. Selected attributes of the solutions in the tabu list (i.e., tabu-active) are marked and solutions that involve these attributes are prohibited. An intermediate-term structure contains intensification rules to drive the search into promising areas of the search space. To this aim, the solutions that involve some predetermined attributes may be induced or banned (e.g., solutions with undesirable values for certain variables may be banned or solutions with desirable values for certain variables may be encouraged), or certain moves may be prevented or stimulated. A long-term memory structure introduces diversification rules that guide the search towards new regions by, for example, resetting the algorithm if the search becomes stuck in a suboptimal region.

7.3.1.4 Evolutionary Computation

Evolutionary computation is a population-based global search approach and subfield of AI to solve optimization problems by using a problem-solving mechanism, based on a simulation of natural evolution processes, such as randomization, natural selection, and mutation. Different extensions and variants of evolutionary computation that basically depend on evolutionary principles can be used to solve specific types of optimization problems involving various data structures, including EAs, evolutionary programming, evolution strategy, differential evolution, memetic algorithms, grammatical evolution, and GAs. EAs are a subset of evolutionary computations and a family of population-based algorithmic trial and error problem-solvers with a metaheuristic character. EA utilizes the characteristics of biological evolution, such as reproduction, natural selection, mutation, recombination, and survival of the fittest. In EA, the potential solutions to an optimization problem are represented by individuals in a population, and the algorithm tries to determine which solutions live and how to create the required diversity in the solution space, by generating an initial set of candidate solutions and updating it in each iteration, using mutation and recombination operators. The solution set is updated by the iterative application of the above operators, and randomly removing worse or less desired solutions, due to an evaluation based on the objective function in the optimization problem being solved (i.e., fitness function), the value corresponding to the solutions, to obtain a set composed of gradually evolved solutions. This procedure is represented by natural selection in biological evolution. Many operators in EA depend on stochastic processes. For example, the selection process can be deterministic or stochastic, in stochastic selection it is more likely to select the solutions with a higher fitness function value. In addition, the recombination and mutation operators randomly choose the elements of solutions to change. EA can generate high-quality solutions to a wide range of problems with less computational effort and in a reasonable time, making them commonly preferred in optimization and search problems. However, the fitness function evaluation may result in computational complexity, which can be dealt with using fitness approximation.

GA is a well-known EA-based metaheuristic method and is commonly used in operations research and computer science. GA replicates the paradigm of biological evolution and utilizes bio-inspired operators including natural selection, mutation, and crossover to produce highly optimized solutions to optimization problems. As in the other EA-based algorithms, GA tries to iteratively evolve a set of candidate solutions, starting from the initial set that is composed of randomly generated individuals, toward better solutions in terms of the fitness function value. The basics of this procedure is inspired by the phenomenon that the fittest individuals in nature have the highest chance to survive. The candidate solutions are called individuals (chromosomes, genotypes, or phenotypes), the set that comprise these solutions is called population and the population in each iteration is called a generation in GA. Solutions (chromosomes) involve encoded information represented by genes, each of which is assigned to a decision variable. GA, recombines, mutates, and alters these genes iteratively to modify the population and form new populations that may contain better solutions. To this aim, the fitness function, corresponding to every individual of the population, is computed and evaluated in each generation. Due to this evaluation, the good (or more fit) individuals are stochastically chosen, and the genotype of these individuals are modified to establish a new generation. This procedure is repeated for the new generation in the next iteration of the algorithm and ends when either the prespecified time limit is over, a satisfactory level for fitness function value has been reached, or a maximum number of generations, also specified by the user, has been produced.

The main prerequirements of a typical GA are; the fitness function and a genetic representation of each candidate solution to facilitate the modification (mutation and crossover) operations, which is generally an array of bits. The genes of each individual in the population are typically represented in binary as strings of 0 and 1 s, however, other encodings are also possible, depending on the type of the problem. A standard genetic representation is showed in Fig. 7.5

Once the prerequisites of GA are defined, the algorithm starts with selecting an initial population of solutions randomly forming the entire search

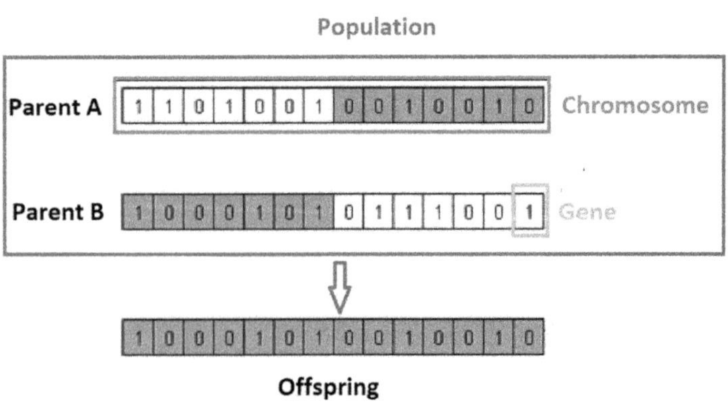

FIGURE 7.5 A standard genetic representation.

space, and continues by iteratively repeating the main operators, mutation, crossover, inversion, and selection, to improve the population. The initial population may be in varying sizes based on the scale and nature of the problem. Large-scale problems may contain thousands of candidate solutions in a population, whereas a few hundred possible solutions may exist in a population of small- and medium-scale problems.

Selection procedure follows the initial population generation. By selection operator, some of the individuals in the current population are chosen considering the fitness function value of each solution. The chosen individuals are called parents and they are used to breed and form the next generation, called offspring or a child. The best solutions are preferably selected by certain selection methods, due to the ratings corresponding to fitness of each solution, which depicts the quality of the solution, whereas some other methods rate the fitness of a randomly selected sample in the population. Both approaches give priority to the fittest individuals in the population and enable the passing of higher quality genes to the next generation, however, the former approach may require extensive computational effort and may be time-consuming. The fitness function depends on the type and target of the problem and may be the same with the objective function of the problem. It may be minimizing the total supply chain cost, minimizing the total transportation distance, or maximizing the total value of objects that fill a knapsack with fixed capacity. Simulation can also be used to calculate the fitness function value of a chromosome, where it is hard to express the fitness function (e.g., mechanical design problems). The selection operator enables the average fitness of the population to be increased since the fittest solutions are selected to breed the next generation. However, a small number of less fit solutions are also selected to ensure the genetic diversity of the creation of offspring solutions.

After the selection procedure, a new generation is created from the selected parents through a combination of genetic operators. Although a range of operators, such as migration, regrouping or colonization–extinction, can be used, two reproduction methods are commonly utilized and known to be the main genetic operators, crossover (also known as recombination), and mutation. The crossover operator produces new chromosomes (individuals), in other words, a child solution, by combining current solutions in pairs. The mutation operator is used to make changes in genes of the chromosomes. Through crossover and mutation, a different population is generated so that the involved solutions share some of the characteristics of their parents. This procedure continues, by selecting new parents from the population created in the current iteration to generate new children, and to establish a new population of solutions of desired size. Although generating new child solutions based on the use of two parents is more common and more suitable to biological systems, some research has shown that using recombination and mutation operators on more than two parents generate higher quality

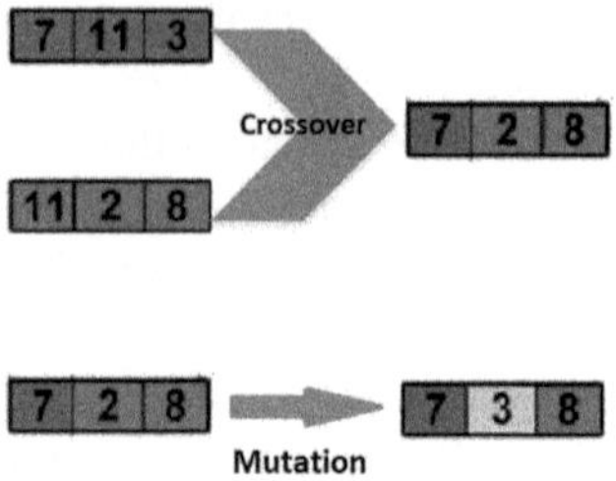

FIGURE 7.6 The crossover and mutation processes.

chromosomes (Eiben et al., 1994). Fig. 7.6 illustrates the crossover and mutation processes.

The above-mentioned evolutionary process is stopped when the termination condition or one of the conditions introduced by the decision-maker is satisfied. Commonly used termination conditions are, a certain number of generations derived or iterations, a specific computation time limit, a good enough solution and fitness value that satisfies the targets, no or insignificant improvements in the fitness value in a specified number of iterations, and certain budget limits.

7.3.1.5 Swarm Intelligence

Swarm intelligence is the study field of bio-inspired AI that replicates the collective behavior and cooperation of large numbers of homogeneous, self-organized, and decentralized agents (e.g., fish, birds, bees, ants, and even bacteria and microbes). These agents follow simple rules and interact with each other in order to solve natural problems including prey evading, colony location, task allocation in colonies, and foraging for food in the natural environment. No centralized control mechanism dictates these rules, instead, they act locally according to an intelligent global behavior with a certain degree of randomness and instinct. Although there are many algorithms that are inspired by the collective behavior of the swarms in nature, two categories of metaheuristics based on swarm intelligence are commonly used to search for the global optimum solution in optimization problems, PSO and ACO.

PSO, inspired by the behavior of a bird flock, is an EA developed by Eberhart and Kermedy (1995). PSO has been extensively used as a computational search and global optimization method, due to its advantage of avoiding being stuck in a local optima, by iteratively trying to improve a candidate solution considering a specified measure of quality. PSO basically solves a problem with a population (called as swarm) of initial candidate solutions (called particles), by moving the particles in the search space, with regards to some mathematical formulations, based on each particle's velocity and instant position in the search space. The movement of each particle is guided by two factors, its best known position and the best known positions discovered by the other particles up to that time in the search space. These positions are updated when improved positions are explored by the other particles. This iterative discovery of the improved positions leads the swarm to move toward the best solutions in the search space. This procedure is repeated until a satisfactory solution to the problem is be discovered.

The PSO algorithm starts with an initial population having random velocities and positions in the search space. Each solution (particle) in this initial set of solutions is evaluated, based on a fitness function, and each particle's position and fitness value is recorded, as well as the fitness value and position of the particle with the best fitness value in the population. Then new velocities and positions are determined for each particle and the fitness value of each particle is reevaluated. If a new position with a better fitness value than the former value is discovered, the position and fitness value is updated for individual particles. Also, if a new position is found with a better fitness value than the fitness value of the best solution in the former population, the best position in the population is updated. Then velocities and positions are updated again. By this repeated search procedure, the positions of the particles are guided to accumulate around the best position in the search space with the best fitness function value. When the termination criteria (the maximum iteration number, the runtime limit, or a satisfactory solution) is met, the search procedure is terminated. One of the advantages of PSO is that, it has a memory ability in that it can keep the position of the best solution in the population as well as the best position of each solution, and pass this information to the other particles and to the next iteration. Also, the convergence rate in the PSO algorithm depends on the particle velocity, hence it has the ability to complete the search process in a shorter time than the metaheuristics. The flow of PSO algorithm is illustrated in Fig. 7.7.

ACO focuses on the probabilistic algorithms for solving optimization problems inspired by the stigmergy and foraging behavior of ant colonies. The basic idea is finding good paths to achieve a solution with desired objective value through graphs. It has been introduced by Dorigo et al. (1996), to search for the best path in a graph, based on the behavior of ants seeking a path between their colony and a source of food. Since then, the algorithm has been improved to solve a wide range of computational problems focusing on the other aspects of the behavior of ants, however, the basic idea remains the same; the artificial ants iteratively move on a graph (i.e., the construction graph) with nodes or edges, to construct good paths to reach the desired objective (finding food for the case of natural ant colonies).

The searching procedure is biased by the pheromone trails in nature, which are created based on the stigmergy behavior—a form of self-organization social network mechanism of indirect coordination, between the agents of intelligent species in nature, of which the basic principle is leaving a trace in the environment, to stimulate and guide the subsequent actions, by

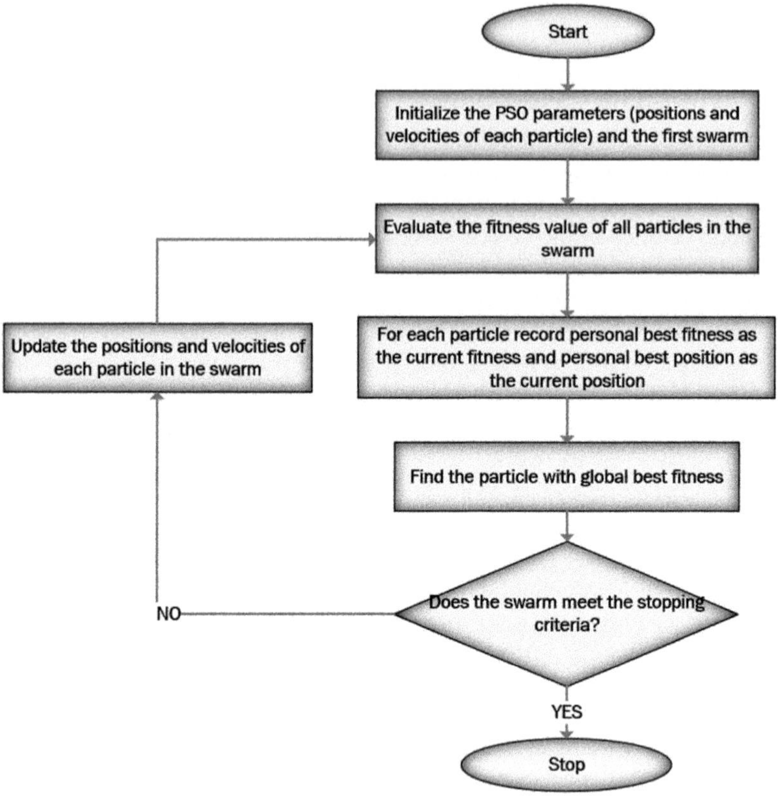

FIGURE 7.7 The flow of PSO algorithm.

the same or a different agent, to systematically accomplish the task of ants. In this scope, the ants of some species initially randomly roam to find food, and once finding the food take it to their colony laying down pheromone trails. When the other members of the colony find this trail, they do not randomly wander to find food anymore, but instead they follow the trail and if they find the food source, reinforce it.

An important point in this behavior is that, after a certain time, the pheromone trail may start to evaporate, which makes it useless for other ants. The length of the trail directly impacts the evaporation time, and a longer trail requires more time for an ant to travel down and back again, which results in a quickly evaporated trail. However, a shorter path is followed more frequently, and thus the pheromone density becomes stronger. As the pheromone intensifies, the trail is preferred by more ants, and positive feedback leads to all ants following a single path to reach the food after a certain time. Evaporation of useless pheromone trails has the advantage of avoiding a convergence to a locally optimal solution. The reason is that the paths followed by the first ants would tend to be extremely attractive to other ants, if there were no evaporation at all, which would further result in a limited exploration of the search space.

ACO has a repetitive structure, which updates the pheromone trails in each iteration with regards to the direction and usage frequency of each path followed by artificial ants to move on the search graph, representing the problem to solve. By following the pheromone trails changed by the artificial ants iteratively, the information to reach the best solutions is kept, passed to the next iteration and updated in the next iteration. First of all, an initial pheromone trail is determined, and each artificial ant is located in a randomly selected edge of the construction graph and starts to move along the edges (or nodes) of the graph visiting the next edge stochastically based on a certain local search probability. The probability of visiting an edge is highly impacted by the pheromone values, so that the higher the pheromone value belonging to an edge, the higher the probability an ant will select that specific edge. Each of the pheromone values is initially decreased by a certain percentage. Each edge then receives an amount of additional pheromone proportional to the quality of the solutions to which it belongs (there is one solution per ant). An ant constructs a solution once it has visited all edges of the graph. Each ant keeps a memory of its path, and hence in subsequent steps it chooses between the edges that do not lead to the other edges that it has already visited previously. Once all the ants have completed their tour, the length of the trail that was followed by each ant is computed and the local pheromone values are updated. The global pheromone value is determined considering all local values obtained in the previous iterations. This procedure is iteratively applied until a stopping criterion is met.

There is a range of extensions to the basic ACO algorithm. The most popular variations include Elitist Ant System (Dorigo et al., 1996), Rank-Based Ant System (Bullnheimer, et al., 1997), MAX−MIN Ant System (Stützle and Hoos, 2000), Hyper-Cube Framework (Blum and Dorigo, 2004), Continuous Orthogonal Ant Colony, and Recursive ACO. Most of them differentiate in the pheromone update stage. For example, *Elitist Ant System* proposes a pheromone update rule, which uses the solutions that are produced in the respective iteration besides the known best solution, for updating the pheromone values. In *Rank-Based Ant System*, all solutions are ranked according to their length and the amount of pheromone deposited. A weighted structure is used to demonstrate that solutions with shorter paths deposit more pheromone than the solutions with longer paths. Hence, the known best solution has the highest impact on the pheromone update at each iteration, while a selection of the best solutions constructed at that current iteration influences the update depending on their ranks (Blum, 2005). *Max−Min Ant System*, which is one of the most powerful ACO variant today, presents maximum and minimum pheromone amounts that are used as lower bound and upper bound to the pheromone values. All edges are initialized to the lower bound, the value of the upper bound is updated each time and a new known best solution is found by the algorithm. The pheromone deposit mechanism of *Continuous Orthogonal Ant Colony* enables ants to search for solutions collaboratively. By using an orthogonal design method, ants in the feasible domain can explore their chosen regions rapidly and efficiently, with enhanced global search capability. *Recursive ACO* is a recursive form of ant system, which divides the whole search domain into several subdomains and solves the objective on these subdomains. The results from all subdomains are compared and the best of them are promoted for the next level. One of the most recent developments is the *Hyper-Cube Framework* for ACO algorithms. Rather than being an ACO variant, the Hyper-Cube Framework is a framework for implementing ACO algorithms. This framework automatically handles the scaling of the objective function values and limits the pheromone values to the interval [0, 1].

There are many other bio-inspired algorithms in the field of swarm intelligence to solve computational problems replicating the collective and cooperative behavior of the colonies in nature, including but not limited to, Bee Algorithms (eg, Honey Bee Algorithm (Tovey, 2004), Artificial Bee Colony Optimization (Karaboga and Basturk, 2007)), Firefly Algorithm (Yang, 2010), Wasp Swarm Algorithm (Pinto et al., 2007), and Bacteria Chemotaxis Algorithm (Muller et al., 2001).

7.3.2 Machine Learning

Machine learning is a subfield of AI dealing with data analysis and analytical model building based on the idea that machines (or systems) can learn from data, without explicit human intervention, to support decision-making in computational problems. Machine learning techniques are able to establish algorithms and models to solve complex problems focusing on making predictions utilizing methods, such as computational statistics, data mining, and mathematical optimization. Machine learning techniques are also useful in learning and forming behavioral profiles for a number of entities to detect meaningful anomalies. A wide range of methods can be utilized, to make computer systems learn from data and construct models, based on these learnings, including but not limited to, statistical learning techniques (e.g., Bayesian networks, k-means clustering, hidden Markov model, nearest neighbor model, regression analysis, factor analysis, and discriminant analysis), artificial neural networks (ANNs), inductive logic programming, instance-based learning, support vector machines (SVMs), information fuzzy networks, decision tree algorithms, and artificial immune system (AIS). Among them, three well-known machine learning method are summarized in the following; ANNs, AISs and SVMs.

ANNs (McCulloch and Pitts, 1943) are among the most popular machine learning techniques to solve complex multivariate nonlinear problems, reliably using appropriate amounts of data and a training algorithm, based on the activities and interactions of interconnected biological neurons in the human brain. The basic ability of ANNs is that it can learn the behaviors behind complicated nonlinear functions, involving statistical data from a training set of data, with the help of artificial neurons (units or nodes) in different layers, connecting each other to transfer the information from one neuron to the others connected to it (a process similar to the synapse between biological neurons to transmit a signal from one neuron to the others). In each layer, a different kind of transformation may be conducted on the inputs and after traversing the layers multiple times, the signals are transferred from the first layer (input layer) to the last layer (output layer). After receiving the information, the artificial neuron processes it and then signals to other artificial neurons. Commonly, the signal transferred from one node to another is a real number, and at each node the value of output is determined based on a nonlinear function of the sum of its inputs. Each neuron and connection has a weight, of which amount is adjusted as the output is generated by the learning procedure. The strength of the signal increases or decreases due to the amount of weight. The next node is triggered only when the weighted sum of input crosses a threshold value, and the output is calculated, as a response of function activation. This self-learning and self-adapting process is conducted with a training algorithm, which provides an architecture for learning the relationships between input and output parameters in the training data set, and establishes an analytical model, based on these learnings to predict the output under the influence of various operational parameters.

ANNs are good alternatives to traditional empirical modeling, based on polynomial and linear regressions and statistical approaches (Kose, 2008). The major advantage of ANNs is that a mathematical model is not necessary to represent the handled phenomena, which significantly decreases the time required for model development. ANNs are also known to be powerful machine learning tools with their ability to fix the values considering an error from the output. There are a number of variants and extensions to ANNs, convolutional neural networks, feed-forward neural networks, large memory storage and retrieval neural networks, radial basis function neural networks, Kohonen self-organizing network, deep stacking networks and deep predictive coding networks. Also, in recent years the combination of ANNs and fuzzy logic has been an attractive research area, termed as Adaptive Neuro Fuzzy Inference System (ANFIS). This integrated computational intelligence method integrates the self-learning ability of the ANNs and the capability of representing the vagueness in linguistic parameters of fuzzy inference. ANFIS provides an efficient tool to capture the dynamic and nonlinear problems with a large amount of data, especially when physical relationships between data are not fully understood, or inaccurate data and data defined by expertise is the case (Nayak et al., 2004). MATLAB, Neural designer, Neuroph, and Darknet are among the most frequently used software to develop and solve ANN frameworks.

AISs comprise a class of bio-inspired computational methods inspired by the functions and characteristics of the immune system (especially memory, detection of anomalies, and learning) in the human body. This rule-based machine learning technique incorporates the metaphors and the perceived properties of the immune system in the development of solutions to engineering problems. The immune system is a vital defense system that intends to protect the body from toxic substances and pathogens, such as microbes and viruses, by producing the immune response when necessary. There are two major tasks that the biological immune system aims to accomplish, the detection and elimination of pathogens. This procedure is accomplished by differentiating between self-cells (the cells that originally belong to the organism) and harmful nonself-cells (harmful cells that do not originally belong to the organism), and pathogens. The nonself-cells and pathogens are targeted to be removed with the help of the defensive layers of the immune system. In a similar way to the biological immune system, the AISs change the structure of an antigen population by adding new antigens that supplant previous antigens in each iteration. The common techniques that AISs use are inspired by the specific immunological functions and mechanisms of the natural immune system. Clonal selection algorithms and negative selection algorithms are among these techniques, each of which can be applied to different kinds of problems. The negative selection algorithm is mostly employed for pattern recognition, anomaly detection, classification, computer and network security, and time series analysis problems, whereas the clonal selection algorithm is suitable for multiobjective and combinatorial optimization. The clonal selection algorithm is based on the clonal selection theory of acquired immunity, which accounts for the behavior and capabilities of antibodies in the acquired immune system (Burnet, 1957). The basic idea of the theory is that given an initial repertoire of general immune cells, the immune system has the capability of adapting the densities and compositions of the cells to protect the organism from environmental pathogenic dangers through combining selection, reproduction, and accumulated variation processes in billions of cells. The system also predicts the properties of the pathogen that the organism will be exposed to before acquiring the required information for the type and timing defense. The negative selection algorithm focuses on the behavior of discrimination of self- and nonself-cells in the acquired immune system, which captures the preparation and continuous maintenance of a set of immune cells except autoimmune cells (or self-reactive immune cells). To this aim, a negative selection procedure is executed to select and remove self-reactive cells during cell creation and proliferation. Inspired by this mechanism, the algorithm suggests that the anticipatory guesses made in clonal selection are filtered to categorize unknown (nonnormal or nonself) data. To this aim, a model of changes, anomalies, or unknown data is built by creating patterns that do not match with known (self or normal) patterns. This model is then utilized to either monitor the current normal or new data by searching for matches with the nonnormal patterns.

SVMs are well-known machine learning models involving learning algorithms that can be used in problems associated with classification and regression. To solve classification problems, as a non-probabilistic binary linear classifier, SVMs construct a model, to assign items to one category or another, given a set of training examples, each marked as belonging to one or the other of two categories. The model comprises each example represented by a point in n-dimensional space (n is the number of features of the items), so that the value of a particular coordinate is the same as the value of each feature. Then, a hyperplane is aimed to be found that differentiates the two classes explicitly and the examples of the separate categories are divided using this hyperplane, which acts as a gap that is as wide as possible. New items are then located into the same space and predictions, associated with the category that the items belong to, are made based on the side of the gap they fall.

Single and hybrid machine learning approaches have been used for decision-making in design and management of biomass-based production chains, especially in forecasting the values of specific problem parameters. Machine learning tools can be used to predict the behavior and performance of nonlinear and complex biomass-based systems and to determine and control variables to improve the operating conditions of a system's environment. AI methods are useful in system modeling involving complicated mappings and system identification, due to the advantages of performing prediction, classification, and generalization rapidly once trained, and not requiring a complete knowledge on process kinetics, which can be extremely difficult to obtain from the highly nonlinear behavior of biological and chemical processes.

The common examples include, but not limited to:

- estimation of methane concentration (or biogas yield) in the biomass mixture in the anaerobic digesters regulating the input values such as temperature, pH, moisture content, and solid content,
- estimation of bio-fuel characteristics, including composition, yield, viscosity, density, water content,
- performance estimation of engines such as electrical efficiency, thermal efficiency, energy consumption, emissions, and exhaust temperature, based on the input parameters, including the compression ratio, load, blend properties, pressure, and injection timing,
- classifying the bio-fuel into different classes according to their origin and basic characteristics,
- estimation of bio-fuel prices in a specific country for specific years,
- controlling the specific parameters of conversion processes (such as the retention time in anaerobic digestion systems), based on the environmental and system conditions (e.g., temperature, thermal conductivity, feeding flow, digester type),

- prediction of combustion efficiency in a boiler, adjusting the values of biomass mixture percentage, moisture content of biomass, excess air, and secondary air during combustion,
- prediction of capital and operational costs associated with different technology investments.

7.4 GEOGRAPHIC INFORMATION SYSTEMS

Geographic Information System (GIS) is an effective tool for collecting, recording, storing, analyzing, presenting, and managing geographic and spatial data using interactive maps. With the help of GIS applications, decision-makers can edit spatial and geographic data in maps, analyze this data by creating interactive searches and visualize the conclusions of these processes. GIS methods and tools are widely used in engineering applications, especially in design, planning, and management of transportation/logistics systems, facility design, specifying and allocation of resource locations (specifically for agricultural resources), and determining the potential regions with high demand. Other areas that depend on collection, analysis and display of spatial data are telecommunications, agriculture, mining, insurance services, and business.

In decision-making for the design, planning, and management of biomass-based production chains, GIS can be used as a supporting tool to display the geographical allotment and spatial distribution of sites for biomass sources, as well as the associated yields, suitable areas that can be specified as candidate sites to construct facilities, transportation distances to determine shortest paths between different nodes in the production chain, and demand regions and densities. This information can then be utilized to parameterize the optimization model.

The main data set for GIS applications consists of spatial data and the key index variable is "location" where the attribute data (additional information about spatial characteristics) is captured in the process with reference to locations on the earth, which are identified using the longitude, latitude, and altitude parameters. For instance, the location of a candidate site for a biomass-based conversion facility is the spatial data, whereas the attribute data may include the distance of this site to biomass source sites, the slope of the area, distance to transportation links and infrastructure, size of the site, distance to water bodies, and distance to residential areas. As another example, the location of an agricultural area for biomass production is the spatial data, and the type and yield of biomass may be the associated attribute data. Relating these two data types enables the user to utilize GIS as an effective problem-solving and decision-making tool for design, planning, and management problems. The attribute data can be visualized and analyzed in a spatial environment to determine the location of features such as candidate sites for facility construction, resource sites, and demand nodes, to be used as parameters in the decision-making procedure. In addition, the relationship of one feature to other features can be identified and the density (or size) of a specific feature in a given location can be determined. Also, by GIS applications the change in a particular area can be observed over time and future conditions can be estimated, to make or modify a strategic decision (e.g., changing the location of a conversion facility or capacity expansion), or a tactical decision (e.g., changing the bio-product distribution route or modifying the production schedule and amounts) on a course of action, or to evaluate the consequences of a certain plan, activity, or policy. For example, the land-use maps are commonly used to demonstrate the expansions in agricultural areas, the level of deforestation, or residential development. There is a wide range of GIS tools and software available throughout the world. Some examples include, ArcGIS, GeoMedia, Mapinfo Professional, SAGA GIS, and GRASS GIS.

7.5 GAME THEORY

Game theory is a study based on mathematical models that deals with conflict and cooperation between independent rational decision-makers within a competitive environment. Game theory is based on the idea of zero-sum games, in which one decision-maker's gains lead to losses for the other actors. The technique is commonly used as a strategy science in economics and business, political science, biology, psychology, philosophy, computer science, and engineering. The "game" concept represents an interactive situation, in which independent participants share some rules and consequences to make rational decisions. These decisions influence a number of economic agents that are in relation with the decision-maker, such as employees, investors, customers, suppliers, and rivals, and the decision-maker must anticipate the reactions of these agents reliably for accurate solutions. At the first stage to apply game theory, information should be gathered or assumptions should be made, associated with the following details; the identity and preferences of rational independent actors, the rules related to the strategic actions that the participants are allowed to perform, the impacts of these strategic actions, and decisions on the outcome of the game. There are two broad categories of game theory, cooperative and noncooperative. Noncooperative game theory deals with how intelligent individuals interact with each other in order to achieve, independently, the optimized value of their individual benefits. In this type of game, only self-forcing individual agreements are allowed where participants cannot form alliances. In cooperative games, the players are supposed to establish a coalition under a set of binding conditions towards optimizing one shared target. The major drawback of cooperative games is that the overall assessment of the payoffs is performed based on cooperative negotiations, without considerable attention to the individual behavior of the companies, related to their reaction to varying cases in a competitively uncertain market environment. This may result in inaccurate assessment of strategies and eventually to suboptimal decision-making.

A game can also be symmetric or asymmetric. In symmetric games, the pay-offs (or utilities, in other words, the numeric representation of punishments or rewards) for playing a certain strategy is dependent on the other strategies, without consideration to the actors playing these strategies, which means that if the actors of the game are changed, there will be no need to change the pay-offs to the strategies. In asymmetric games, the players do not have the identical strategy repertoires. In game theory, actors can either play their game simultaneously or sequentially. In simultaneous games, both participants make a move simultaneously and the following actors are not aware of the previous actions of the former players, whereas in sequential games (also called as dynamic games) the latter actors have knowledge about earlier actions. In sequential games, the information about every earlier action may be perfect or imperfect. A game can also be either a zero-sum-game or nonzero-sum game. In zero-sum-games, the amount gained by one actor is only at the equal loss of others, in other words, the total gain to all actors in the game is always zero for every combination of strategies. This allows deducting the gain of one player from the losses of other players. In nonzero-sum-games, the benefit to one player is not the same amount as the loss of the others. Nash Equilibrium (Nash, 1951) and Stackelberg games can solve noncooperative games. Nash Equilibrium is an appropriate approach when the game is simultaneous and symmetric, whereas the Stackelberg method can be employed in case of a conflict of interests between different actors in asymmetric games.

To start the game, after listing the identity of each participant and the alternative actions (or strategies or reactions) available for each participant, a matrix is formed that involves the actions of the first player in the rows and the actions of the second player in the columns. Each cell in the matrix contains two numbers to represent the payoff values associated with the actions of each player. Prisoner's Dilemma game, which is one of the most popular games, is given as example to illustrate the procedure. In this game, two players are calculated to be caught by the police, who represent the partners in crime, and each one is suggested to confess their crime in separate jails where they have been placed individually. The following matrix represents the game:

	Not confess	Confess
Not confess	(4 , 4) Cell 1	(-2 , 8) Cell 3
Confess	(8 , −2) Cell 2	(2 , 2) Cell 4

The higher the numbers in each cell, the more utility is associated with each strategy. After the game is defined, each players' strategies are tried, to assess predicting the behavior of each player in the next step. Varying numbers can be used each time the game is played. Because some parameters may change drastically if the game is repeated or in case of any future interaction between the players. However, some basic assumptions remain the same, for example, the main aim of both players is to rationally maximize their own utility from the game. The above matrix shows that:

- If both players choose to not confess (cell 1), they share the acquisition of their crime, which is represented by 6 units of pay-off for each player, and they both go free.
- If one player confesses and the other does not, player 1 gets the highest score of 8 units of pay-off and goes free as he has testified against the other. The player who has not confessed goes to prison, which results in the lowest value of pay-off of -2 (cells 2 and 3).
- If both players confess their crime (cell 4), both are sent to prison but with a reduced term. This situation is represented by giving each one 2 units of pay-off. This case is better than having the other player confess, but not so good as going free.

This game provides a simplified representation of situations with significant impacts. Different economic, environmental, or social strategies in various problems can be evaluated instead of confess/not confess phenomena. For example, "setting a high price" and "setting a low price" strategies may be considered for a specific product, in comparison with the prices of the same product in the associated market, with the primary aim of obtaining a high revenue from the sales of the product. In this game, the players may be two companies with conflicting and competing objectives. High prices may be good for both companies, however, it may be the best for each individual company to set a low price, while the other sets a high price.

In management of biomass-based production chains, each utility value can be calculated, based on the objective function of the problem. The objective function is called the "pay-off function" in case of a maximization objective (e.g., maximizing the profit), or "loss function" in case of a minimization objective (e.g., minimizing the cost). Some examples of the application of game theory to biomass-based production chain management problems can be given as follows:

- Distribution planning for bio-fuel products to determine the types, amounts, and allocation decisions in a cooperative and multiagent environment under binding protocols.
- Planning of a decentralized biomass supply chain containing producer and consumer by a cooperative model, based on contracts with binding conditions about matching production amounts with the consumer demand.

- Optimizing the biomass production levels and sale prices associated with different competitive suppliers in a supply chain.
- Optimizing the conflicting objectives between a producer (e.g., maximizing the revenue from sales) and a consumer (e.g., minimizing the purchasing cost) in a supply chain using a noncooperative game model.
- Optimizing the purchasing and inventory amount in a supply chain with a single producer and a single consumer using a noncooperative game, based on the shortage penalty, where the producer has to pay an allowance in case of any shortage in the supply per unit of product that has not been fulfilled.

7.6 DISCUSSION

In recent years, studies on optimization of biomass-based production chains has concentrated on the development and application of advanced modeling and optimization methods and tools to make enhanced design, planning and management decisions. Linear and mixed integer programming models have been extensively used to make strategic, tactical, and operational decisions by the decision-makers to solve design and management problems in oil and chemical industries in the last decades. However, the current research show that the use of hybrid methodologies, that integrate the advantages and features of two or more methodologies in one decision-making framework, for the optimization of biomass-based production chains, results in more efficient decisions, especially for complex problems. For example, the combination of GIS and a modeling/optimization technique, such as mathematical modeling or simulation, enables us to parametrize the decision models reliably. In addition, the existence of uncertainty in biomass-based production chains, as well as the computational complexities and challenges in modeling all complicated characteristics and interactions in production and distribution systems, restrict the decision-making abilities of standalone modeling approaches. These challenges increase in large-scale systems that occupy a large geographical area, multiple time periods in the planning horizon, the combination of different stages in supply chains and longer planning horizons. Hybrid modeling and optimization approaches can capture the uncertainty and handle the computational complexity of large-scale network problems. The development of algorithmic solution approaches and advances in computational software enable us to handle a large number of decision variables, parameters, and constraints along with multiple objectives, in tandem. However, tractability of the model is an important consideration to keep the computational complexity and times within acceptable limits. For this reason, the simplicity and ease of implementation are crucial factors in model construction for practitioners. Decomposition methods and certain algorithms can also be utilized to deal with the complex structure of large-size problems and reduce the solution time.

Although problem-specific parameters and variables are involved, most of the optimization models can be employed to solve problems in various biomass-based production chains, given that some common assumptions are made that are valid for all of them and data are available, however, some modifications in the parameters, variables, constraints, and objective functions may be needed to fit the model in the problem requirements.

REFERENCES

Achterberg, T., Koch, T., Martin, A., 2005. Branching rules revisited. Operat. Res. Lett. 33, 42–54.

Bai, Y., Hwang, T., Kang, S., Ouyang, Y., 2011. Bio-fuel refinery location and supply chain planning under traffic congestion. Transport. Res. Part B: Methodol. 45 (1), 162–175.

Blum, C., 2005. Ant colony optimization: Introduction and recent trends. Phys. Life Rev. 2 (4), 353–373.

Blum, C., Dorigo, M., 2004. The hyper-cube framework for ant colony optimization. IEEE Trans. Systems Man Cybernetics—Part B 34 (2), 1161–1172.

Bullnheimer, B., Hartl, R.F., Strauß, C., 1997. A new rank-based version of the Ant System. A computational study. In: Working Papers SFB "Adaptive Information Systems and Modeling in Economics and Management Science", WU Vienna University of Economics and Business, Vienna.

Burnet, F.M., 1957. A modification of Jerne's theory of antibody production using the concept of clonal selection. Aust. J. Sci. 20, 67–69.

Dantzig, G.B., 1990. Origins of the simplex method. In: A History of Scientific Computing, Reading, MA, pp. 141–151.

Dantzig, G., Thapa, M., 1997. Linear programming 1: introduction (Springer Series in Operations Research and Financial Engineering).

Dorigo, M., Maniezzo, V., Colorni, A., 1996. Ant system: optimization by a colony of cooperating agents. IEEE Trans. Systems Man Cybernetics—Part B 26 (1), 29–41.

Eberhart, R.C., Kermedy, J., 1995. New optimizer using particles Swarm theory. In: Proceedings of the Sixth International Symposium on Micro Machine and Hmm Science, Nagoya, Japan.

Eiben, A.E., Raué, P.E., Ruttkay, Z., 1994. Genetic algorithms with multi-parent recombination. In: Davidor, Y., Schwefel, H.P., Männer, R. (Eds.), Parallel Problem Solving From Nature — PPSN III. PPSN 1994. Lecture Notes in Computer Science. Springer, Berlin, Heidelberg, p. 866.

Feo, T.A., Resende, M.G.C., 1989. A probabilistic heuristic for a computationally difficult set covering problem. Operat. Res. Lett. 8 (2), 67–71.

Gunnarsson, H., Rönnqvist, M., Lundgren, J.T., 2004. Supply chain modeling of forest fuel. European Journal of Operational Research 158 (1), 103–123.

Ivanov, D., Sokolov, B., Kaeschel, J., 2010. A multi-structural framework for adaptive supply chain planning and operations control with structure dynamics considerations. Europ. J. Operat. Res. 200 (2), 409–420.

Karaboga, D., Basturk, B., 2007. Artificial bee colony (ABC) optimization algorithm for solving constrained optimization problems. In: Melin, P., Castillo, O., Aguilar, L.T., Kacprzyk, J., Pedrycz, W. (Eds.), Foundations of Fuzzy Logic and Soft Computing. Lecture Notes in Computer Science. Springer, Berlin, Heidelberg, p. 4529.

Karmarkar, N., 1984. A new polynomial-time algorithm for linear programming. In: Proceedings of the Sixteenth Annual ACM Symposium on Theory of Computing, Washington, DC, pp. 302–311.

Kose, E., 2008. Modeling of color perception of different age groups using artificial neural networks. Exp. Syst. Applicat. 34 (3), 2129–2139.

Li, X., Ouyang, Y., 2010. A continuum approximation approach to reliable facility location design under correlated probabilistic disruptions. Transport. Res. Part B: Methodol. 44 (4), 535–548.

Marufuzzaman, M., Eksioglu, S.D., Li, X., Wang, J., 2014. Analyzing the impact of intermodal-related risk to the design and management of bio-fuel supply chain. Transport. Res. Part E: Logist. Transport. Rev. 69, 122–145.

McCulloch, W.S., Pitts, W., 1943. A logical calculus of the ideas immanent in nervous activity. Bull. Math. Biophys. 5 (4), 115–133.

Mol, R.M., Jogems, M.A.H., Van Beek, P., Gigler, J.K., 1997. Simulation and optimization of the logistics of biomass fuel collection. NJAS Wageningen J. Life Sci. 45 (1), 217–228.

Muller, S.D., Marchetto, J., Airaghi, S., Koumoutsakos, P., 2001. Optimization based on bacterial chemotaxis. IEEE Transact. Evolu. Comput. 6, 16–29.

Nash, J., 1951. Non-Cooperative Games. Ann. Math. 54 (2), 286–295.

Nayak, P.C., Sudheer, K.P., Rangan, D.M., Ramasastri, K.S., 2004. A neuro-fuzzy computing technique for modeling hydrological time series. J. Hydrol. 291 (1-2), 52–66.

Nilsson, D., 1999a. SHAM—A simulation model for designing straw fuel delivery systems. Part I. Model Descrip. Biomass Bioenergy 16, 25–38.

Nilsson, D., 1999b. SHAM—A simulation model for designing straw fuel delivery systems. Part II. Model Applicat. Biomass Bioenergy 16, 39–50.

Pinto, P.C., Runkler, T.A., Sousa, J.M.C., 2007. Wasp swarm algorithm for dynamic MAX-SAT problems. In: Beliczynski, B., Dzielinski, A., Iwanowski, M., Ribeiro, B. (Eds.), Adaptive and Natural Computing Algorithms. Lecture Notes in Computer Science, vol. 4431. Springer, Berlin, Heidelberg.

Shabani, N., Sowlati, T., 2013. A mixed integer non-linear programming model for tactical value chain optimization of a wood biomass power plant. Appl. Energy 104, 353–361.

Sokhansanj, S., Kumar, A., Turhollow, A.F., 2006. Development and implementation of integrated biomass supply analysis and logistics model (IBSAL). Biomass Bioenergy 30 (10), 838–847.

Stützle, T., Hoos, H.H., 2000. MAX–MIN ant system. Fut. Generat. Comp. Syst. 16 (8), 889–914.

Tovey, C., 2004. The honey bee algorithm: a biological inspired approach to internet server optimization. The alumni magazine for ISyE at Georgia Institute of technology, 13–15.

Yang, X.S., 2010. Firefly algorithm, stochastic test functions and design optimization. Int. J. Bio-Inspired Comput. 2, 78–84.

Yılmaz Balaman, S., 2014. A Fuzzy Goal Programming-Based Decision Support System for Design and Management of Biomass to Energy Supply Chains. Ph.D. Thesis, Dokuz Eylül University, İzmir.

FURTHER READING

Ting, C.K., 2005. On the mean convergence time of multi-parent genetic algorithms without selection. In: Capcarrère, M.S., Freitas, A.A., Bentley, P.J., Johnson, C.G., Timmis, J. (Eds.), Advances in Artificial Life. ECAL 2005. Lecture Notes in Computer Science. Springer, Berlin, Heidelberg, p. 3630.

Index

Note: Page numbers followed by "*f*" and "*t*" refer to figures and tables, respectively.

A

Ablative system, 38–39
ACO. *See* Ant colony optimization (ACO)
Adaptive Neuro Fuzzy Inference System (ANFIS), 227
Aerifying, 58
Agent-based simulation, 209
Agricultural/agriculture, 7
  crops, 46–47
  processes, 15
  residues, 8–9
AHP. *See* Analytical hierarchy process (AHP)
AI. *See* Artificial intelligence (AI)
AIS. *See* Artificial immune system (AIS)
Algae, 19–20
  production, 117
Ambient storages, 58
Ambiguity, 131
Anaerobic
  digestion, 13, 41–46, 45*f*
  enzymatic conversion process, 46
  fermentation, 14–15
  microorganisms, 31
Analytic network process (ANP), 166
Analytical hierarchy process (AHP), 166
ANFIS. *See* Adaptive Neuro Fuzzy Inference System (ANFIS)
Animal wastes, 13
Annealing process, 216
ANNs. *See* Artificial neural networks (ANNs)
ANP. *See* Analytic network process (ANP)
Ant colony optimization (ACO), 214
  recursive, 225–226
Aquatic Species Program, 19–20
Artificial immune system (AIS), 226
Artificial intelligence (AI), 213
  AI-based approaches, 213–230
    computational search and optimization methods, 213–226
    machine learning, 226–230
Artificial neural networks (ANNs), 226
Ash, content and composition of, 29–30
Asymmetric games, 231–232
Attached growth digesters, 44–46

B

Bacterial biodegradation process, 42
Bark, 10–11
Batch reactor, 41
Bellman and Zadeh's "min operator", 177
Benders' decomposition, 126, 201–202
Benefit–cost ratio, 101–103
Bio-char, 36
Bio-chemical conversion
  systems, 54
  technologies, 41–48
    anaerobic digestion, 42–46
    fermentation, 46–48
Bio-chemical degradation, 16
Bio-chemical/biological processes, 25
Bio-crude. *See* Crude-like oil
Bio-diesel, 50, 67–68, 71, 89
Bio-energy, 13
  conversion processes, 13
  production, 119–120
Bio-fertilizer production, 13
Bio-fuel, 13, 72, 115
  consumption, 67–72
  handling, 71–72
  production system, 105–107
  storage, 67–72
  transportation, 67–72
Bio-hydrogen, 51–52
Bio-oil. *See* Crude-like oil
Bio-products, 25
Biodegradability, 42
Biodiversity loss, 85–86
Bioethanol, 71–72

Biogas, 68, 70
Biomass, 1–19, *2f*, 26, *26f*, 161
  dedicated energy crops, 2–7
  distribution system sustainability, 61–62
  extraction, 49–51
  feedstock, 67, 119–120
  nonrenewable-based production systems, 20–21
  production, 19–20
  residues, 8–12
  slurry, 60–61
  sources, 58, 149–150
  storage, 58
  sustainability certification system, 98–99
  uncertainty in supply of, 116–117
  wastes, 13–19
Biomass-based energy production, 62
Biomass-based production chains, 56, 59, 114, 150, 152–153
  artificial intelligence-based approaches, 213–230
  decision levels in design and management, 143–153
    operational decisions, 151–153
    strategic decisions, 144–149
    tactical decisions, 149–151
  decision-making in design and management, 153–178
  game theory, 231–234
  GIS, 230–231
  handling uncertainties in, 121–139
    approaches for uncertainty modeling, 122–137
    methods for uncertainty impact evaluation, 137–139
    scenario analyses, 138–139
    sensitivity analysis, 137–138
  mathematical modeling-based approaches, 185–206
  simulation-based approaches, 206–212
  sustainability, 77–90
    assessment in design and management, 100–105
    economic sustainability, 78–82
    environmental sustainability, 82–86
    pillars of sustainability and corresponding factors, *79f*
    social sustainability, 86–89
    technical sustainability, 89–90
  uncertainty
    in governmental and regulatory policies, 119
    in market demand and prices, 114–116
    natural uncertainties and disruption possibility, 120–121
    sources in, 113–121
    in supply of biomass and materials, 116–117
    technology and process uncertainties, 117–119
    in transportation and logistics, 119–120
Biomass-based production systems
  industry, 119
  LCA of, 105–108, *106f*
  non-renewable resources, 53–54
  systems, technologies, and processes, 25–51
    biochemical conversion technologies, 41–48
    biomass extraction, 49–51
    thermochemical conversion technologies, 26–41
  technologies, 51–53
Biomass-based supply chains and logistics networks
  bio-fuel storage, transportation, and consumption, 67–72
  biomass conversion, 67
  biomass cultivation, harvesting, collection, and storage, 58–59
  biomass pretreatment, 59
  biomass transportation, 59–67
  energy production and transmission, 72–73
  process flow in biomass-based energy production chain, *57f*
  structures and key components, 55–73
Biomethane, 43–44, 68, 70, 72
  liquefying, 70–71
Biorenewable resources, 1
Blue water footprint, 159–160
Branch and bound method, 203–206, *206f*, *207t*
Bundling, 12
Butanol, 48

C

Cake, 50
Carbon
  balance, 83–84
  dioxide, 27–28
  emissions, 83–84
  footprint, 83–84, 159
CBA. *See* Cost-benefit analysis (CBA)

CEA. *See* Cost-effectiveness analysis (CEA)
Cell-free enzymatic process, 52
Cellulose, 35–36
Centralized systems, 58
Centrifugation, 47–48
Certification systems, 98
Chance-constrained programming, 128–129
Chemical extraction. *See* Solvent extraction
Chemical Oxygen Demand, 160–161
Chipping, 12
CHP units. *See* Combined heat and power units (CHP units)
Circulating fluidized beds, 38–39
furnaces, 29
Climate change, 2–3
Clonal selection theory, 228
Co-current gasifier. *See* Downdraft gasifier
Co-modality, 65
Coal or biomass, 29
Coding system, 71–72
Column generation, 199–201
Combined heat and power units (CHP units), 43, 63
Combustion, 14, 26–29
Comminution systems, 12
Competitiveness, 80
Completely mixed reactors. *See* Continuous stirred tank reactors (CSTR)
Composting process, 15–16
Computational search and optimization methods, 213–226
evolutionary computation, 219–222
greedy algorithms, 215–216
SA, 216–218
swarm intelligence, 222–226
TS, 218–219
Computer simulation models, 206–208, 210
Container trailers, 63–64
Continuous Orthogonal Ant Colony, 225–226
Continuous simulation, 209
Continuous stirred tank reactors (CSTR), 41
Convection, 48
Conversion
processes, 1, 13, 33
systems, 25
Cooling process, 216
Corn, 49
Cost-benefit analysis (CBA), 101–103
Cost-effectiveness analysis (CEA), 101–103
Counter-current gasifier. *See* Updraft gasifier
Crop management, 8–9
Crop residue, 8
Cross-draft gasifiers, 31–33
Crossover operator, 221–222
Crude-like oil, 40
CSTR. *See* Continuous stirred tank reactors (CSTR)
Cutting plane method, 201–202
Cutting process, 201
Cyanobacteria, 19–20
Cycling, 198
Cyclonic systems, 38–39

D

Dairy manure, 13
Dantzig Wolfe decomposition, 199–201
Data envelopment analysis (DEA), 169
Decentralized storage, 58–59
Decision variables, 185–186, 191–192
Decision-making in design and management, 153–178
economic objectives and criteria, 154–158
environmental objectives and criteria, 158–161
multiple attribute and multiple criteria decision-making methods, 165–178
social objectives and criteria, 161–163
technical criteria, 163–164
Decision-making trial and evaluation laboratory (DEMATEL), 168–169
Decomposition methods, 204
Dedicated energy crops, 2–7
herbaceous energy crops, 3
plantation and harvesting, 4–6
transportation of, 61–67
treatment and storage, 7
woody energy crops, 3–4
Degeneracy, 198
Demand management, 115–116
Demand-oriented strategies, 80–81
DEMATEL. *See* Decision-making trial and evaluation laboratory (DEMATEL)
Dense phase reactors, 31–33
Density factor, 31
DES. *See* Discrete event simulation (DES)
Design constraints, 193–194
Deterministic-based parameters, 122
Deviational variables, 171–172
Diffusion, 48
Digestion, 14–15
Discrete event simulation (DES), 209
Dissolved-air flotation method, 18
District heating, 72–73

Downdraft gasifier, 31–33
Downstream stage, 105–107
Drying stage, 30
Dynamic games, 231–232
Dynamic simulation models, 209–210

E

EAs. *See* Evolutionary algorithms (EAs)
Eco-indicator 99, 107–108
Ecology-based LCA (Eco-based LCA), 105–107
Economic assessment, 101–103
Economic objectives and criteria, 154–158
Economic sustainability, 78–82
Elementary algebra, 198–199
Elephant grass (e-grass). *See* Miscanthus
ELimination Et Choix Traduisant la Réalité or Elimination and Choice Translating Reality (ELECTRE), 167–168
Elitist Ant System, 225–226
Employment effect, total, 86–87
Energy. *See also* Biomass
  balance, 193–194
  conversion process, 8
  crops, 3, 26–27
  efficiency, 81
  footprint, 160
  production and transmission, 72–73
Energy Policy Act (2005), 71
Enterprise resource planning (ERP), 152–153
Entrained-flow gasifiers, 33
Environmental assessment, 103–104
Environmental objectives and criteria, 158–161
Environmental sustainability, 82–86
ε-constraint method, 170–171
ERP. *See* Enterprise resource planning (ERP)
Ethanol, 31, 68, 70, 89
Ethanol–gasoline blends, 70
European Commission, 99
European Committee, 8
Evolutionary algorithms (EAs), 214
Evolutionary computation, 219–222
Excreta, 15–16
Exergy, 193–194
Expeller press, 50–51
Extraction, 12, 47–48
Extruder or screw press. *See* Expeller press

F

Fecal sludge. *See* Excreta
Feeding rate, 43
Feedstock, 26–27, 29
Fermentation, 41–42, 46–48
FGP. *See* Fuzzy goal programming (FGP)
Filtration, 47–48
Financial flow, 55
Fischer-Tropsch synthesis, 53
  coal-derived syngas, 31
Fixed beds, 38–39
Fixed film digesters, 44–46
Fixed trucks, 63–64
Fixed-bed
  combustion systems, 28–29
  reactor, 41
Flames, 27–28
Floating gas holders, 68–69
Fluidized beds, 38–39
  combustion systems, 28–29
  gasifiers, 33
Food
  and beverage processing, 16
  vs. fuel debate, 83
  production, 2–3, 8
  security, 89, 162–163
Forest biomass, 85
Forest ecosystem health, 85
Forestry residues, 9–11
Fossil fuel-based energy production systems, 25
Fossil sources-based production technologies, 51
FST. *See* Fuzzy Set Theory (FST)
Fuel production, 2–3
Functional unit, 105–107
Fuzzy decision-making, 131–137
Fuzzy goal programming (FGP), 174
Fuzzy linear programming, 134
Fuzzy multiple objective linear programming, 172–178
  pay-off table, 175*t*
Fuzzy optimization, 131–132
Fuzzy probabilistic programming, 140–141
Fuzzy Set Theory (FST), 132–133

G

GAs. *See* Genetic algorithms (GAs)
Gasification, 14–15, 29–33, 32*f*
Gasifier, 29–30, 34*t*
Gasohol, 71–72
Gasoline, 71–72
Gasoline gallon equivalent, 105–107
Genetic algorithms (GAs), 214, 219–220
Geographic information system (GIS), 179, 230–231

GFN. *See* Global Footprint Network (GFN)
GHGs. *See* Greenhouse gases (GHGs)
GIS. *See* Geographic information system (GIS)
Global Footprint Network (GFN), 160
Global warming potential (GWP), 107–108
  indicator, 159
Goal programming technique, 171–172
Good Agricultural Practices (GlobalGAP), 99–100
Governmental and regulatory policies, uncertainty in, 119
GRASP. *See* Greedy randomized adaptive search (GRASP)
Grass, 7
Grassy or forage energy crops. *See* Herbaceous energy crops
Gray water footprint, 159–160
Greedy algorithms, 215–216
Greedy randomized adaptive search (GRASP), 214–216
Green water footprint, 159–160
Greenhouse gases (GHGs), 158–159
  emissions, 36, 82–84, 189
Grid connection, 72
GWP. *See* Global warming potential (GWP)

H

HCR. *See* Heavy crude (HCR)
Heat distribution system, 73
Heat transfer, 26–27
Heavy crude (HCR), 40
Hemicellulose, 35–36
Herbaceous energy crops, 3, 5
Heuristic approaches, 194, 213
Hill climbing algorithms, 214
Homogeneity, 42
HRT. *See* Hydraulic retention time (HRT)
HTL. *See* Hydrothermal liquefaction (HTL)
Husbandry processes, 15
Hybrid machine learning approaches, 229
Hydraulic press, 50–51
Hydraulic retention time (HRT), 44–46
Hydrothermal liquefaction (HTL), 26, 39–41
Hyper-Cube Framework, 225–226
Hypethral storages, 58

I

ILP. *See* Integer Linear Programming (ILP)
Immune system, 228
IMPACT 2002 +, 107–108
Indices, 190
Industrial wastes, 16–17
Information flow, 55
Integer Linear Programming (ILP), 187
Integer programming model (IP model), 186
Interior point methods, 194–199
Intermodal transportation, 61, 65–66
Internal rate of return (IRR), 101–103, 154–155, 157
International Energy Agency, 62
Interval-uncertainty modeling, 130
Inventory analysis, 107–108
Inventory management, 56
  phases, 150–151
IP model. *See* Integer programming model (IP model)
IRR. *See* Internal rate of return (IRR)
Itaconic acid, 48

J

Job creation in biomass-based production chains, 86–87

L

L-shaped method. *See* Benders' decomposition
"Lagrange Multiplier", 202–203
Lagrangian relaxation, 202–203
Land
  footprint, 83, 161
  land-use, 83
  management, 83
  price, 62
  transformations, 85–86
Large capacity production systems, 65
Large-scale conversion systems, 62–63
Larger scale biomass systems, 63
LCA. *See* Life cycle assessment (LCA)
LCCA. *See* Life Cycle Cost Analysis (LCCA)
LCI. *See* Life cycle inventory (LCI)
LCIA. *See* Life cycle impact assessment (LCIA)
LCR. *See* Light crude (LCR)
Lean phase gasifiers, 33
Life cycle assessment (LCA), 104
  of biomass-based production systems, 105–108, 106*f*
Life Cycle Cost Analysis (LCCA), 108
Life cycle impact assessment (LCIA), 107–108
Life cycle inventory (LCI), 107

Light crude (LCR), 40
Lignin content, 42
Lignocellulosic biomass, 46–47
Lignocellulosic materials, 8, 26–27
Linear membership function, 173, 176
Linear programming model (LP model), 186–192
  decision variables, 191–192
  indices, 190
  methods and tools to solve NLP problems and, 194–206
    branch and bound, 203–206
    column generation and Dantzig Wolfe decomposition, 199–201
    cutting plane method and benders' decomposition, 201–202
    Lagrangian relaxation, 202–203
    simplex method and interior point methods, 194–199
Liquid
  bio-oil, 36–37
  liquid/gaseous bio-fuels, 60–61
Liquid Petroleum Gas, 31
Logic-based methods, 213
Logistics
  network, 65
  uncertainty in, 119–120
Low-pressure internal systems, 68–69
LP model. *See* Linear programming model (LP model)

M

Machine learning, 226–230
Macroalgae, 19–20
MADM. *See* Multiple attribute decision-making (MADM)
Maritime transportation, 60
Market demand and prices, uncertainty in, 114–116
Material flow, 55
Material requirement planning (MRP), 149
Mathematical modeling-based approaches, 185–206
  LP, 186–192
  methods and tools to solve LP and NLP problems, 194–206
  NLP, 192–194
Mathematical programming model, 185–186
MAUA. *See* Multiple attribute utility analysis (MAUA)
Max–Min Ant System, 225–226
MCA. *See* Multicriteria analysis (MCA)
MCS. *See* Monte Carlo simulation (MCS)
Mechanical expression, 49–50
Mechanical pressing. *See* Mechanical expression
Metaheuristics algorithms, 214
Microalgae, 19–20
Microbes, 51, 67–68
MILP. *See* Mixed Integer Linear Programming (MILP)
MIP model. *See* Mixed integer programming model (MIP model)
Miscanthus, 3
Mixed Integer Linear Programming (MILP), 187
Mixed integer programming model (MIP model), 186
MODM. *See* Multiple objective decision-making (MODM)
Moisture content, 21, 29–30, 36
MOLP. *See* Multiple objective linear programming (MOLP)
Monte Carlo simulation (MCS), 209
MRP. *See* Material requirement planning (MRP)
MSW. *See* Municipal solid waste (MSW)
Multi-stage stochastic programming, 126
Multicriteria analysis (MCA), 101–103
Multiobjective optimization, 110
Multiple attribute and multiple criteria decision-making methods, 165–178
Multiple attribute decision-making (MADM), 153–154, 165–170
Multiple attribute utility analysis (MAUA), 169
Multiple objective decision-making (MODM), 153–154, 170–178
Multiple objective linear programming (MOLP), 172
Municipal solid waste (MSW), 14–15

N

Nash Equilibrium, 231–232
Natural uncertainties and disruption possibility, 120–121
Net present value (NPV), 101–103, 154–157
NLP model. *See* Nonlinear programming model (NLP model)
Non-carbon GHGs, 83–84
Non-preemptive goal programming, 172
Non-renewable resources, 53–54

Noncooperative game theory, 231
Nonfood crops, 2–3
Nonlinear mass balance, 193–194
Nonlinear programming model (NLP model), 186, 192–194
  methods and tools to solve LP and, 194–206
Nonrenewable-based production systems, 20–21, 116
NPV. *See* Net present value (NPV)

O

Odor, 13
Oleiferous seeds, 49–50
Operational decisions, 151–153
Optimization, 185
  methods, 213–226
Organic loading rate, 44–46
Organization, Rangement Et Synthese De Donnes Relationnelles method (ORESTE method), 167–168
Outsourcing, 147–148
Oxidation, 30

P

Particle size, 36
Particle swarm optimization (PSO), 214, 224*f*
Pay-off function, 233–234
Pelletizing, 26–27
PFRs. *See* Plug flow reactors (PFRs)
Phosphorus (P), 13
Pipeline system, 66–67
Pipeline transportation, 60–61, 64–65
Pivot variable, 196
Plantation and harvesting of dedicated energy crops
  planting and harvesting, 5–6
  site selection and preplanting site preparation, 4–5
Plasma gasification technology, 52
Plug flow reactors (PFRs), 41
Poultry litter, 13
Preemptive goal programming, 172
Preference Ranking Organization Method for Enrichment Evaluations (PROMETHEE), 167–168
Preplanting site preparation, 4–5
Pretreatment
  biomass, 59
  method, 59
Price uncertainty, 116
Price volatility, 82
Primal–dual interior point method, 198–199
Primary forestry residues, 9–10
Prisoner's Dilemma game, 232
Probabilistic parameters, 122
Production planning, 56, 150
Production technologies, 89–90
Productivity, 81
Profitability, 80
Progressive hedging algorithm, 126
PROMETHEE. *See* Preference Ranking Organization Method for Enrichment Evaluations (PROMETHEE)
PSO. *See* Particle swarm optimization (PSO)
Pure Integer Linear Programming (Pure ILP), 187
Purification, 47–48
Pyrolysis, 14–15, 30, 33–39

Q

Quadratic programming, 193

R

Rail transportation, 60, 64
Rank-Based Ant System, 225–226
Rapeseed, 49
Raw biomass, 59
RDF. *See* Refuse-derived fuel (RDF)
Real-world phenomena, 185
Recursive ACO, 225–226
Recursive stochastic programming, 128
Red Tractor, 99–100
Reduction phase, 31
Refuse-derived fuel (RDF), 17
Renewable basis, 1
Renewable Energy Directive, 99–100
Renewable energy production, 2–3
Residues, 8–12
  agricultural, 8–9
  of fermentation process, 44
  forestry, 9–11
  harvesting, treatment and storage, 11–12
  transportation, 61–67
Resilience, 80–81
Responsiveness, 80–81
Retention time, 43
Return on investment (ROI), 154–155
Rhizomes. *See* Underground storage organs
Road
  haulage, 60
  transportation, 60, 63

Robust optimization, 129–131
Robustness, 129
ROI. *See* Return on investment (ROI)
Round Table on Responsible Soy EU RED (RTRS EU RED), 99–100
Rural development, contribution to, 87–88

S

SA. *See* Simulated annealing (SA)
Safflower, 49
Salvage value, 164
SAN. *See* Sustainable Agriculture Network (SAN)
SAW approach. *See* Simple Additive Weighting approach (SAW approach)
Sawdust, 10–11
Scoring method. *See* Simple Additive Weighting approach (SAW approach)
Secondary forestry residues, 9–10
Seed, 5
Sensitivity analysis, 137–138
Service level-oriented strategies, 80–81
Sewage sludge, 15–16
Short-Rotation Coppice, 3–4
Short-rotation woody crops. *See* Woody energy crops
Simple Additive Weighting approach (SAW approach), 168
Simplex method, 194–199, 195*f*
Simulated annealing (SA), 214, 216–218, 217*f*
Simulation-based approaches, 206–212, 212*t*
Single machine learning approaches, 229
Site selection and preplanting site preparation, 4–5
SLCA. *See* Social Life Cycle Assessment (SLCA)
Sludge, 18
Sludge retention time (SRT), 44–46
Social acceptability, 88
Social assessment, 104–105
Social Life Cycle Assessment (SLCA), 108
Social objectives and criteria, 161–163
Social return on investment (SROI), 163
Social sustainability, 86–89
  standards, 98
Soil contamination, 85
Soil degradation, 85
Solid wastes, 19
Solvent extraction, 49–50
Soybean, 49
SRC, 4–6
SROI. *See* Social return on investment (SROI)
SRT. *See* Sludge retention time (SRT)
Stabilization, 18
Standard genetic representation, 220*f*
Steel tanks, 68
Stochastic dynamic programming, 128
Stochastic programming, 122–129
Storage containers, 68
Storage tank, 67–68
Strategic decisions, 144–149
Straw retention, 11–12
Succinic acid, 48
Sunflower, 49
Supply chain, 55. *See also* Biomass-based supply chains and logistics networks
  management, 55–56
  nodes, 187–188
  total supply chain profit maximization, 191*b*
Support vector machines (SVMs), 226
Suspended growth digesters, 44–46
Sustainability
  assessment in design and management, 100–105
  of biomass distribution system, 61–62
  of biomass-based production chains, 77–90
  criteria, 91–92
  indicators, 93*t*
    standards, and certification schemes, 91–100
  LCA of biomass-based production systems, 105–108, 106*f*
Sustainable Agriculture Network (SAN), 99–100
SVMs. *See* Support vector machines (SVMs)
Swarm intelligence, 222–226
Sweeping gas flow rate, 37
Switchgrass, 3
Symmetric games, 231–232
Syngas, 31
  fermentation, 31
System boundary, 105–107
System dynamics, 209–210

T

Tabu search (TS), 214, 216–218
Tactical decisions, 149–151
Tannery operations, 17
Tar content, 29–30
Technical sustainability, 89–90

Technique for Order Preference by Similarity to Ideal Solution (TOPSIS), 167
Thermal crackling. *See* Pyrolysis
Thermal energy storage, 73
Thermochemical conversion technologies, 26–41, 54
combustion, 26–29
gasification, 29–33
HTL, 39–41
pyrolysis, 33–39
Thermochemical liquefaction. *See* Hydrothermal liquefaction (HTL)
Thermochemical processes, 25
Timber or pulp production, 9–10
"Tipping fee", 14
TOPSIS. *See* Technique for Order Preference by Similarity to Ideal Solution (TOPSIS)
Total water footprint, 84–85
Tractor–trailers, 63–64
Transesterification, 50
Transportation, 12
bio-fuel, 67–72
biomass, 59–67
of dedicated energy crops, residues, and wastes, 61–67
and distribution planning, 56
intermodal, 61, 65–66
maritime, 60
pipeline, 60–61, 64–65
rail, 60, 64
residues, 61–67
road, 60
uncertainty in, 119–120
wastes, 61–67
waterborne, 64
waterway, 60
Trapezoidal membership function, 133–134, 135*f*
Triangular membership function, 133, 134*f*
TS. *See* Tabu search (TS)
Tube trailers, 70–71
Turbulence, 28

U

Uncertainty
in governmental and regulatory policies, 119
handling uncertainties in biomass-based production chains, 121–139
in market demand and prices, 114–116
natural uncertainties and disruption possibility, 120–121
sources in biomass-based production chains, 113–121
in supply of biomass and materials, 116–117
technology and process uncertainties, 117–119
in transportation and logistics, 119–120
Underground storage organs, 5
Updraft gasifier, 31–33
Upstream stage, 105–107

V

VIseKriterijumsa Optimizacija I Kompromisno Resenje (VIKOR), 166–167

W

Waste-to-energy technologies, 14
Wastes
animal wastes, 13
industrial wastes, 16–17
MSW, 14–15
transportation, 61–67
treatment and storage of wastes, 17–19
wastewater and sewage sludge, 15–16
Wastewater and sewage sludge, 15–16
Water
footprint, 159–160
vapor, 27–28
Water-soluble inorganic compounds, 40
Water-use, 84–85
Waterborne transportation, 64
Waterway transportation, 60
Weighed additive approach, 178
Weighted linear combination. *See* Simple Additive Weighting approach (SAW approach)
Werners' "fuzzy and" operator, 177
Wood pellets, 10–11
Woody biomass-based systems, 88
Woody energy crops, 3–4

CPI Antony Rowe
Chippenham, UK
2018-10-11 20:10